EXPLAINING LIFE
THROUGH EVOLUTION

EXPLAINING LIFE
THROUGH EVOLUTION

PROSANTA CHAKRABARTY

THE MIT PRESS   CAMBRIDGE, MASSACHUSETTS   LONDON, ENGLAND

First published by Penguin Random House India.

The MIT Press would like to thank the anonymous peer reviewers who provided comments on drafts of this book. The generous work of academic experts is essential for establishing the authority and quality of our publications. We acknowledge with gratitude the contributions of these otherwise uncredited readers.

This book was set in ITC Stone and Avenir by New Best-set Typesetters Ltd. Printed and bound in the United States of America.

Library of Congress Cataloging-in-Publication Data

Names: Chakrabarty, Prosanta, author.
Title: Explaining life through evolution / Prosanta Chakrabarty.
Description: Cambridge, Massachusetts : The MIT Press, [2023] | "First published by Penguin Random House India."—Title page verso. | Includes bibliographical references and index.
Identifiers: LCCN 2022052104 (print) | LCCN 2022052105 (ebook) | ISBN 9780262546256 (paperback) | ISBN 9780262375467 (pdf) | ISBN 9780262375474 (epub)
Subjects: LCSH: Evolution (Biology) | Natural selection. | Life—Origin.
Classification: LCC QH366.2 .C43115 2023 (print) | LCC QH366.2 (ebook) | DDC 576.8—dc23/eng/20221229
LC record available at https://lccn.loc.gov/2022052104
LC ebook record available at https://lccn.loc.gov/2022052105

10 9 8 7 6 5 4 3 2 1

Dedicated to teachers dedicated to teaching . . .

CONTENTS

PART IV: WHY UNDERSTANDING EVOLUTION MATTERS

I

A PERSONAL PROLOGUE

CALLED TO ACTION

In 2008, the year I moved to Louisiana, the state legislature passed the Louisiana Science Education Act with the backing of Governor Bobby Jindal. The name of the law is misleading since it empowered schools in Louisiana to teach the biblical account of creation in science classes as an alternative to evolution.[1] Jindal went as far as to say that local schools should determine how science is taught in classrooms.[2] That meant that if the local public school's science teacher wanted to teach that all living land animals were the descendants of the animals carried on Noah's ark, then, well, that was just fine with the governor and the supporters of the law.

At the time, I presumed that Bobby Jindal (born "Piyush Jindal") was probably a smart person who was somehow misled. I jokingly called him "my Uncle Bobby" because, as an Indian myself, I could claim all other Indians as kin, and maybe saying the governor was a relative could get me out of a speeding ticket (it couldn't). My Uncle Bobby should have known better than to back antievolution legislation and sign it into law. He was, after all, someone who had a biology degree from Brown University and was a Rhodes Scholar (but going to an Ivy League school and Oxford doesn't necessarily make you smart, just educated). Part of the reason he promoted this law was to pander to the so-called "Religious Right." At the very least, he knew that the law was unconstitutional because of the "separation of church and state" clauses of the First Amendment of the U.S. Constitution. He did know this, and he ignored it. This new attempt at subverting the teaching of evolution was based on a previous Louisiana law, struck down by the Supreme Court in 1987 in *Edwards v. Aguillard.*

So, what's the big deal anyway? Well, you just don't teach your own religious views in a public science class. Science is about observing and testing natural phenomena in order to give reasoned, evidence-based explanations for those events. Religion, on the other hand, can provide some folks with answers to questions science doesn't cover (e.g., What is the meaning of life?) but it can also provide answers that can't always be tested. For instance, let's say your answer to why apples drop to the ground when they fall out of a tree is "God made it happen"; that isn't something I can prove false, because I can't test it. There isn't room for questioning things or scientific inquiry if you believe flatly that "God controls everything that happens."

The other problem with teaching religion in a science class is that there are many religions with a variety of beliefs. Faith-based beliefs about creation differ depending on your religious persuasion. In one version of the Hindu creation myth, the Earth was part of the lotus flower that grew from the navel of Vishnu, and then the world was populated by Brahma and will be destroyed by Shiva.[3] If I taught that version of creation as the truth in my science class, I wouldn't last very long as a teacher.

The problem again with teaching "God controls everything" is that you can't prove or disprove it. In order to respond to a "Noah's flood" scenario, I could say, "There is no boat that can fit a pair of all living land animals, and having just a pair of each species wouldn't provide enough genetic variability to restock the Earth; also it can't rain for forty days and forty nights worldwide because we have a limited amount of water in Earth's atmosphere at any given time; and where did all that extra rainwater go afterward? And there is no evidence of a huge worldwide flood from a few thousand years ago . . ." The reply could always just be, "God did it, so anything is possible." Well, I guess that would be the end of the conversation. "God did it" can be your go-to answer, and that's fine. I'm not here to judge or tell you how to live. The problem arises when that explanation is used to shut down scientific discourse, especially in an academic setting.

Science is all about subjecting explanations of natural phenomena to rigorous testing to see if they hold up. If you can explain everything away with "God did it and controls everything," what's the point of the testing? What's the point of school? What's the point of curiosity?

Religions ask for your faith in explaining the untestable and unprovable; science is here to help explain the rest by making testable predictions. Does science require some faith? Well, you certainly have to have faith that the results and observations you make are based in reality and that the results are the outcomes of predictable phenomena. (This idea of having "faith" in reality dates back before Aristotle but is explicit in the writings of religious philosophers like Albertus Magnus, Thomas Aquinas, Roger Bacon, Avicenna,

Averroës, and surely many others in every culture, religion, and region.) If a deity changed and influenced every scientific experiment on a whim so that we couldn't make predictions and the laws of physics were not immutable, then, frankly, we wouldn't be able to explain our universe with any scientific investigations. But that clearly isn't the case. We scientists can have faith in the fact that what we observe is an unmanipulated reality because our inquiries yield consistent and reliable results that allow us to reveal the secrets of our planet and of our universe ever more exactly. To paraphrase Galileo, "Religion shows the way to go to heaven, not the way the heavens go."

Some would argue that there is room for studying nature in the slightly different religious take of "God started everything and made the rules." That is, you can study the rules of gravity or evolution, for example, and their consequences, without having to determine the origin of those rules. This playbook suits the "non-overlapping magisteria" Stephen Jay Gould describes in his book *Rocks of Ages*.[4] In that book, Gould argues that science and religion can complement each other without interfering with each other. Or, as I would put it more plainly: "Science should not be antireligion; and religion should not be antiscience." There is certainly room for compromise and improved "cultural competency" on both sides.[5]

I don't mean to pick on Louisiana by pointing out that the antievolution law enacted back in 2008 is still on the books, I genuinely love the state and am proud to call it my adopted home. My children were born in Baton Rouge, and I've lived here for more than ten years. Louisiana is far from the only state with issues over the teaching of evolution. In

Alabama, biology textbooks sometimes still include a disclaimer sticker warning you about the "theory of evolution." [6] In 2018, Arizona briefly attempted to have all references to the word "evolution" deleted from the state's science education standards.[7] It isn't just people in the so-called conservative states who object to the teaching of evolution; I've met antievolution people in liberal Ann Arbor (where I received my PhD in evolutionary biology) and New York City (where I grew up). In fact, one in five high school teachers in the United States teach creationism along with evolution in their classes (many teach it as the preferred scientific alternative).[8] It isn't just in the United States either—only 26 percent of people in Afghanistan accept human evolution,[9] and, in 2017, Turkey removed references to evolution from its textbooks.[10] Notably, the United States ranks near Turkey (recently renamed Türkiye) in terms of how human evolution is accepted by the general public.[11]

I'm here with no other motive but to explain the scientific facts that are available. The science is why the vast majority of people who understand those facts accept evolution. I'm not here to challenge your beliefs. We all have our own truths. We get to pick our own beliefs—but we *don't* get to pick our own facts. I'm here to help you better understand how science explains the origin of our species and of all life on Earth.

In discussions with many people who actually accepted evolution, I came to realize that a good percentage of them accepted it as fact because they trust science, even though they didn't really understand the science supporting it. They may understand even less than some people who *don't* accept evolution because those people don't trust science or may see some conflict with their religion. I want to help more people on both sides to understand evolution, so that it might become less divisive. It is ultimately up to you to accept scientific facts (or not), and to convince yourself of whatever "truth" you are willing to acknowledge.

Evolution is not the easiest topic to understand—nor the hardest either ("It ain't rocket surgery," as they say), although it is easy to misinterpret, and it is often portrayed incorrectly. For instance, evolution is often depicted as something like the *March to Progress* (*Road to* Homo sapiens) mural by Rudolph Franz Zallinger (figure 1A), but that depiction is very wrong. This image is often interpreted as presenting evolution as a progressive force making life more "complex" or "perfect," and that we (humans) are the most complex and perfect form of life—we are not—and there is no way to objectively measure "complexity" or "perfection" in any event. Evolution isn't goal oriented or progressive. This mischaracterization led one Louisiana state senator, when told of a long-term evolution experiment on bacteria, to ask, "They evolved into a person?"[12] That experiment, actually called the "long-term evolution experiment" (LTEE), provided a surprising example of an evolutionary adaptation observed as it occurred.[13] In his lab at Michigan State University, Richard Lenski and his colleagues found that some *E. coli* populations (from over 70,000 generations), all captive reared from the same source population, had evolved an adaptive trait.[14*]

---

* Read more about that remarkable LTEE experiment from the many Lenski lab publications. Sadly, the LTEE experiment was paused for the first time after thirty years due to the COVID-19 pandemic; it has now transitioned to a new lab at the University of Texas.

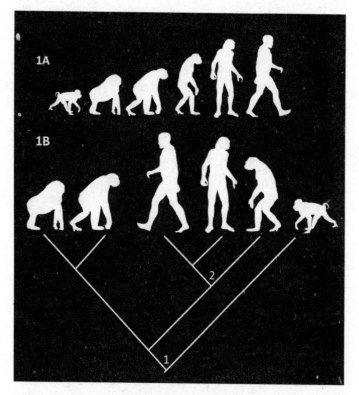

Figure 1 Image 1A—the most widely used image for depicting evolution—is wrong. It gives the impression to many that we transformed from a monkey to a chimp and then to a hunched "caveman" in a direct series leading to "modern man." In reality, we share common ancestry with all of these forms, but we did not progress from them; rather, we progressed *with* them from our shared ancestor. Depictions like image 1A, based on the 1941 mural *March of Progress* (original title *Road to* Homo sapiens) by Rudolph Franz Zallinger, have done more harm than good to those trying to better explain evolution.

A more accurate way of depicting evolution is with a phylogenetic tree (an evolutionary branching diagram showing the links between closest relatives). The branches split in two, and those pairs sharing a "node" (the point where the branches connect) are more closely related to each other. Here the tailed monkey is the closest relative to—not the "ancestor" of—the five other primates. The primates' ancestors are represented in the nodes. Node 1 is the common ancestor of the six primates (including humans) depicted above, node 2 is the common ancestor of chimps + modern humans + extinct human species.

That is very clear evidence of evolution (i.e., change in the inherited characteristics of a population). What the state senator believed, based on a common misunderstanding of evolution, was that the only definitive evidence of evolution from the Lenski experiment would be if the bacteria actually turned into humans. Even given an infinite number of generations, however, those bacteria would never turn into humans—that's not how evolution works.

This book is meant to explain how evolution does work (figure 1B) to a broad audience—indeed, to anyone with a mind open to learning. It is also meant to aid those who themselves want to explain the science of evolution to others. Finally, I want to explain here why understanding evolution should matter to everyone in this age of genetic editing and increasingly politicized science.

PULLED INTO THE FIGHT

After that 2008 antievolution law was enacted, many scientists in Louisiana decided to do something about it. Barbara Forrest, a philosophy professor at Southeastern Louisiana University, led the charge among academics to fight the good fight against the creationist movement that was trying to set up antievolution laws all over the country (see her coauthored book *Creationism's Trojan Horse*[1]). The best-known spokesperson against the creationists was a Louisiana high school student at the time, Zack Kopplin.[2] If Forrest was exposing the Trojan horse for what it was, Zack was Hector, the young prince out to take down Achilles. Zack made national news by making articulate arguments defending evolution and science. At Louisiana State University (LSU), my colleagues Christopher Austin, Bryan Carstens, and others worked hard to make Evolution a core course for anyone majoring in biology. A core course means all biology majors are required to take it, something that Bobby Jindal didn't have to do at Brown. Perhaps if Bobby had learned evolutionary biology in college, he wouldn't have been moved to put the antievolution law in place. After Evolution became a core course, I decided to step up and teach it myself. The fight of my colleagues became my fight, and I thank them for letting me join the cause. We now teach evolution to several hundred students a semester at LSU.

Despite teaching a class on evolution and sometimes being asked to explain the science of evolution on panels or seminars, I never thought of myself as an advocate for science education. I didn't see the need—I thought that most people who were antievolution were just poorly informed, or so I assumed. It took me a while to realize that I wasn't in a position to make that assumption. While prepping for a talk focused on evolution for the 2018 TED conference, I came to recognize that I was usually speaking to other biologists or biology majors about evolution, that is, to people who already had a fairly good understanding of the fundamentals of biology, but the people helping prep my talk, though experts in public speaking and other subjects, were not biologists. That was probably why I didn't think evolution was such a complicated subject to explain, but I was wrong. It only seemed like an intuitive subject to me because of how long I had been studying and thinking about it.

The early versions of my public evolution talks were often very jokey with lots of why-the-human-body-stinks jokes (e.g., in insult-comic Don Rickles's voice: "Yeah, hang your baby-making dangly bits on the outside of your body between your legs, where they will be easy to bump against a table—good idea!") but I soon realized (with the help of more mature people) that insulting the human body wasn't the best way to explain why we aren't the pinnacle of evolution (but read on—examples ahead). Instead, I decided to focus on how some of the most typical depictions of evolution are misleading and how so many people lack a fundamental understanding of what evolution is and where the facts and theories about it came from. I want to make the science of evolution more accessible, and that's why I wanted to write this book—to help people decide for themselves.

My own family helped to get me thinking about evolution from a different perspective than just as a mild-mannered professor. I have identical twins: every evolutionary biologist should have identical twins. My eleven-year-olds are very different from each other. I think often about how my twins' identical genomes are reflected in their bodies and behaviors. I'm glad my wife kept me from carrying out my evolution experiments of nature versus nurture on our twins (you know, only sending one to school, only hugging the other, and so on). Instead, we are doing a "common garden" experiment, raising them in the same environmental conditions and seeing how they still end up different and distinct.* Obviously, I'm joking . . . sort of. But if I was pushed into science advocacy, it was for them. I want them to grow up in a more scientifically literate society.

Since I'm asking you to be interested in the science of evolution, it may be worth explaining how I first became interested. Bob Carroll at McGill University in Montreal first

---

* I do have a hypothesis that identical twins raised together actively try to be different from each other, whereas those raised apart grow up to be more similar in dress and disposition (look up the "Jim Twins" or "Three Identical Strangers," for example). A sort of "character displacement" (where two similar species living close to each other accentuate their differences) takes place with twins raised together. By chance, we had a second set of identical twins with us at the time of this writing—three-year-old boys who we were fostering—and who only reinforced my hypothesis.

introduced me to the topic—at least to the fun macroevolution bits like dinosaurs and mass extinctions. His funny and inspiring lectures were a big part of my wanting to teach evolutionary biology myself one day. Sadly, Bob died of COVID-19-related complications in 2020. But one of the guest speakers he brought in while I was at McGill was Stephen Jay Gould, perhaps the most influential evolutionary biologist of the last fifty years. And learning that Gould was raised in Bayside, Queens, like myself, probably played a significant role in my continued interest in evolutionary biology.

My first hands-on work in the science of evolution was related to describing a new species of fish when I was an undergrad at the American Museum of Natural History in New York. Under the tutelage of Melanie Stiassny, I not only learned about taxonomy but also developed a love of anatomy and ichthyology. In grad school in a department of evolutionary biologists at the University of Michigan, I was inspired daily to study evolution and the evolutionary relationships between groups of species on the Tree of Life ("phylogenetic systematics"). Douglas Futuyma—who, as I joke, "literally wrote the book on evolution"—because he wrote *Evolution*, the most important undergraduate textbook on evolutionary biology—helped me think about the nitty-gritty of the study of life on Earth, particularly at the microevolutionary, population level.[3] I was also lucky enough to be a teaching assistant for a theme semester at the University of Michigan called "Explore Evolution," when I heard and met with some of the best evolutionary biology speakers. These included Svante Pääbo, who won a Nobel Prize in 2022 for sequencing the Neanderthal genome and who also figured out that genes putting you at greater risk of severe COVID-19 symptoms were inherited from Neanderthals.[4] Richard Lewontin,[5]* who brought a mathematical approach to population genetic theory, and Eugenie Scott, who exposed the antievolution movement's hidden agendas.[6] Being able to help teach the science of evolution to undergrads that first time was eye opening. Few of those undergrads had any exposure to the subject at that point, and they had few resources to help them learn about it on their own. I was privileged in having had those resources and mentors.

Now at LSU, I understand the impact of explaining evolution not just each time I teach my class but whenever I get an email or social media comment about my public talks. (I'm sure I'll get a few more after this book comes out.) I've been called some pretty horrid things by some antievolution folks who don't care to understand the science—and who also don't understand that I'm just presenting the best available knowledge on the subject and trying to make evolution a little more approachable for everyone.

---

* Rest in peace to Richard Lewontin, who died in 2021, along with his Harvard colleague and nemesis E. O. Wilson. Unfortunately, in 2022, we learned that Wilson was not-so-secretly aiding and supporting eugenicists, which reinforced the idea that Wilson's book *Sociobiology*, which caused the rift between him and Lewontin, had some roots in racist ideology.

ON TRUST

The first time I was taught evolution I was in eighth grade, at Marie Curie Middle School 158 in Queens, New York, when I was about thirteen years old. The teacher said, "I'm going to talk about evolution for a moment, I apologize in advance to those of you who are religious . . ." I don't actually remember what he said next, something about human fossils from Africa, but that was it. Why the disclaimer? Was he saying something false? Unproven? Was it different from the earth science he was teaching us that was about continents and glaciers? If I could go back in time, I would raise my hand and ask, "Does every science teacher have to apologize for teaching me evolution?" Except I wouldn't. I loved that teacher, and I know that he and most other science teachers aren't trained to teach evolution, and that many people, even biology majors, aren't exposed to it.[1]

Now I'm the teacher, and I know all too well why people approach the subject of evolution with trepidation—they fear confronting zealots or offending someone's religious views. I occasionally get strange looks from people when I tell them what I teach in Louisiana. People sometimes comment that they have visions of kids throwing holy water on me, screaming at me while I read with my head down from a piece of paper behind some protective glass (a common but false stereotype people have of teaching in "the South"; the glass barrier was just protection from COVID). But I always tell them, "No, no—it isn't like that." I've taught hundreds of students at LSU, and I've never had an aggressive interaction. Students are generally very respectful and come to learn. They might not always like what I say, they might question what I say (I encourage that), but they haven't ever been rude or mean about it. Even in the anonymous evaluations at the end of the term, I haven't had any comments about the science (most negative evaluations are about how my exams and assignments are too hard). I think the thing that helps me avoid those confrontational situations with my students is my saying, "Hey, look, I won't ever be telling you what to believe. I know I can't make you believe something. I can just show you the available scientific facts, show you how we gather and build up the data underlying those facts, and then you have to make your own judgment calls on what you accept or don't. But what I'm teaching, and what I expect you to learn, are the facts scientists have rigorously tested and accept."

If some folks don't believe the moon landings happened, showing them videos of the landings or telling them that you can still see the reflectors and other things the astronauts left behind won't convince them. I can provide all that "proof," but it is still up to them whether to believe it. If they have squarely convinced themselves that the moon landings weren't real, then they have a belief system of mistrust about the people telling them that the landings really happened. Instead, they trust the antiscience people who tell them the landings were faked. That's the core of it: they don't trust science. For folks like that, you have to start far away from the target subject to get them to understand the fundamental science first. In that way, they can see how (and why) the conclusions were drawn, and hopefully that will help them to trust science in general.

So, who do you trust? Ultimately, we trust ourselves and our network of friends, family, and colleagues who have earned our trust. And, yes, trust has to be earned. We trust them until we are given a reason not to. As Hemingway said, "The best way to find out if you can trust somebody is to trust them." I ask you, dear reader, to trust me, but I also ask you to challenge what I write. If there is information that you don't trust here, find out more where you can, from the sources I cite and from sources that you find on your own. I only ask that you examine what information you have trusted in the past carefully.

The most important part of understanding a subject is trust. If you don't trust the source, you won't believe the information on that subject. I can't make you understand evolution without a bit of trust. How can I make you trust me? I can't. I can only pledge to you that when I teach evolution, and when I write about the subject in this book, I am doing my best to explain the most accurate scientific information that we have on the subject. I can only give you the best-tested explanations that scientists like myself have found to explain life on Earth, and I can show you how we got to those explanations.

FACTS AND TRUTH

Now that I've earned your trust—or at least your willingness to read on a bit—let me tell you that, frankly, I don't know the "truth." Or, at least, I can't say with absolute certainty that something is true ("???!!!" I hear you "think scream"). I won't lie to you, but I will admit when there is uncertainty involved. And when it comes to trust, should you trust people who tell you "with absolute certainty" that they know what is really "true"? Of course, acknowledging the inherent uncertainty of knowledge doesn't stop us scientists from pursuing the truth: we know the truth is out there.

So how do you find out if something is true? Well, you can start by proving that it is at least not false. As a scientist, I test hypotheses (which are explanations that answer questions), trying to prove them false. Can you call what is not-false "true"? Well, maybe. As long as you know that some hypotheses are just "not false yet," or "not false in this situation." And some hypotheses hold up under such rigorous testing that we don't question them anymore. We call these "facts." Scientific facts are time-tested explanations of how the world works: for example, an apple drops to the ground when it falls from a tree because of the force of gravity. By continued testing and questioning, we accumulate more and more positive evidence to support a hypothesis. That's why scientists don't question biological evolution: evolution happened, and the evidence is all around us. Evolution is a fact. You can't disprove evolution—there is too much evidence that living things give rise to other living things, which are modified by time and the environment.

We live in an age where many of us think we can get our own "facts" through Google searches on our smartphones. Most people believe whatever they want to believe. They see a social media post that falsely claims that vaccines kill thousands of people every year, and they share it, and their friends share it. They get the sense that this claim is well researched because they see the same meme several times, and it makes sense to them within their circle of like-minded friends.[1] It feels like the truth to them because it seems like everyone else is blind and stupid and not paying attention. But we shouldn't get our facts from social media alone (or maybe, not at all). There we often only find the watered-down, quickly digestible versions of facts (if not something that is outright false with the intent to mislead or misdirect) because it is becoming harder for most of us to find the actual primary sources of facts—where the information actually was derived and reported. Scientific facts are the corroborated results of tested hypotheses. These scientific facts can be found originally from the primary peer-reviewed scientific literature. One easy way is to use Google Scholar instead of Google (go to scholar.google.com—it's still Google, just refined and focused on science journal articles), and find papers with a "Methods" and "Results" section (that means the authors tested something and got an answer). Peer-reviewed articles also include reviews and comment pieces that may come up with novel ideas. But if you are looking for how someone figured out that Greenland sharks can live over 400 years[2] or that the oldest fossils are 3.5 billion years old[3] or that $E = mc^2$,[4] then you need to look at the original papers (the primary sources) that discovered and described those facts.* These papers are not easy to read—even scientists often have difficulty reading other scientists' papers. But you don't need to understand every word of these articles; you only need to get the gist of what the results are, what it means, and why it matters. Much of that can be found in the abstracts of the papers or in their introduction and discussion sections. And you may be surprised how much you enjoy the detective work of finding the sources of scientific facts.

In this book, I try to reference the primary scientific literature for every key fact that I present so that you can find the original research yourself and see how that fact came to be "a fact." The purpose of this book is to wrap some facts about evolution in a neat package to make them a little more digestible and to present them with some background and history.

Some people believe only what they can see, so all of history before they were born is lost to them, and they are blind to all that is beyond their field of vision. Some people never wash their hands "because they don't believe in germs since

---

* Walk into any public university library, and the staff will help you find primary resources in stacks of journal articles in bound volumes. For evolutionary biology, check out the journals *Evolution*, *Systematic Biology*, *Nature*, *Science*, and many others that are also online. Again, I recommend you start by searching in "Google Scholar," https://scholar.google.com/.

germs don't exist if you can't see them."[5] Others may feel obliged to build a homemade rocket to see if there really is any curvature to the Earth, which they believe to be flat.[6] Well, good for them. Basically, they are skeptics, just like the real scientists. The difference is they don't trust the people with microscopes and laboratories who found the microbes that make you sick or the people who could demonstrably show that the Earth is round (even with people flying all "around" it right now). Some people need to find their own truth and don't trust anything they can't see or touch—no scientists or doctors are going to tell them what is true and what is not. I don't think you would have read this far if you were as lost as they are. I hope you are willing to learn at least how evolutionary biologists have come to their explanations. Yes, you'll have to trust me a bit, but we can start slow. Let's get into some history and see how some of those facts about evolution were discovered.

II

THE EVOLUTION REVOLUTION

INTRODUCTION TO EVOLUTION

Look at your hand.

Why do you have five fingers?

Why not ten, or twenty, or one? Why do so many animals have five fingers? Five seems to be the perfect number for most hands. Oddly, the first vertebrates to come onto land had many more fingers—seven or eight.[1] These earliest land vertebrates were fish crawling out to land, at first just briefly, then steadily, progressively, for longer periods of time. As the descendants of that lineage became permanently established on land, the number of fingers decreased to five or fewer and stayed that way.[2]

How about hair? Ever wonder why it is where it is on our bodies? Or why we have hair at all? Well, first off, we're mammals—each with a belly button and a love of dairy and our mothers (well, most of us at least). But we're different from other mammals. Why do we stand upright?

Sure, a kangaroo is a mammal that stands upright, but it has a nice thick tail to balance on, and we don't. "What about a bird, like an ostrich?" you ask. They have beautifully and efficiently fused bones, like the tarsometatarsus between their ankles and their feet, for strength and stability, while we putter around with thirty-three joints in our feet along with about a fourth of all the bones in our body.[3] What were we thinking? Becoming bipedal doesn't seem to have been such a good idea. Doesn't your back, neck, and feet hurt after you've been standing around for a while? Why are our backs curved worse than a busted fender on a rally car, instead of being straight, like a rod? And why do we have to balance our giant heads on top of our curvy spines—seems like a bad idea. Being warm-blooded comes in handy when it's cold, but it makes us need to eat all the time. I bet you're hungry right now, unlike that cold-blooded python chilling for months without having to snack. And nipples on males—what's up with that? Evolution, that's what's up.

Indeed, that's the thing about evolution: you don't get to start from scratch each time, redesigning the body for each new model (species). You must build on what came before. And so sometimes you have to take a nice straight, loosely connected fish spine and stick it together and prop it up with a few extra twists and turns.[4]

Every time I look at the human body, especially my own (and especially when sitting nearly naked waiting in the doctor's office), I think of the ways I would fix it. I think of our aquatic ancestors, the animals that first developed a vertebral column (your "spine"), lungs, fingers, a big, compartmentalized brain and heart, and so on. I think of how much more comfortable we would be in the water, not fighting against gravity, just floating like a happy little sea turtle beneath the waves. So many of our body parts first evolved for use in water. I think of the gill structures we had, that you can still see in a developing human fetus, that transform into your voice box ("larynx") and jaws and into the little bones that allow you to hear.[5] I think of the muscles that help lift the gills in a shark being the same ones that are the muscles of your neck and upper back.[6]

We didn't get our muscles from sharks, nor did we evolve from sharks, but we do share a common aquatic ancestor a long, long time ago before bony and cartilaginous animals diverged. We had a more recent ancestor that was a bony fish, making all of its descendants, like us, technically bony fishes, too. All of us bony fishes—trout, seahorses, donkeys, and humans—are more closely related to one another than any of us are to sharks. Cartilaginous fishes and bony fishes have been evolving independently for a long time (although newly discovered fossil evidence suggests cartilaginous fishes may just be bony fishes that lost bone[7]), but cartilaginous fishes remained relatively stable in their body types (mostly "shark-like" or flattened, as in the skates and rays) than members of the bony fish lineage. Some members of that bony fish lineage stayed in the water and continue to evolve to this day (as part of the over 35,000 species of bony fishes living); others left the water and gave rise to all of the land vertebrates or "tetrapods," of which there are also about 35,000 species (figure 2). I sometimes wonder what it would have been like if cartilaginous fishes had invaded land, too. Just think about it—land sharks! But perhaps the cartilaginous skeleton of a shark isn't strong enough to withstand the force of gravity for long (imagine standing on your floppy cartilaginous ears), even though sharks do have more than twice as much muscle as we do (some 75% versus only about 30% for humans).[8] If our aquatic ancestors

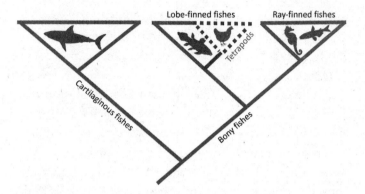

Figure 2 The Vertebrate Tree of Life is the same as the Fish Tree of Life (since we can call all vertebrates "fish" because of our shared ancestry); four-limbed vertebrates ("tetrapods") are in the dashed area of the tree (represented by a hen, but we humans belong in there, too—along with other mammals and with birds, reptiles, and amphibians) as members of the bony fish lineage; specifically, we are lobe-finned fish (in which lineage lungfishes and coelacanths also belong). We could call everything in this tree a "fish" because they all have gills and fins at some stage, even though the animals in the dashed portion of the tree are mainly found on land and have modified gills and fins, as we do. (See plate 1.)

didn't have strong bony skeletons (and other organ and sensory systems adapted to survive in shallow waters), they wouldn't have been able to invade land, and we wouldn't be here. Sometimes people refer to certain animals, especially humans, as "perfectly designed," but our aquatic ancestors had to twist and stretch and rework what they already had. You can't get to the perfect solution for being a land animal from that fishy starting point (as those of us who have ever had backaches or had food stuck down our throats can confirm).

We share a common ancestor with all life on Earth, but we share common ancestors with some lineages of the Tree of

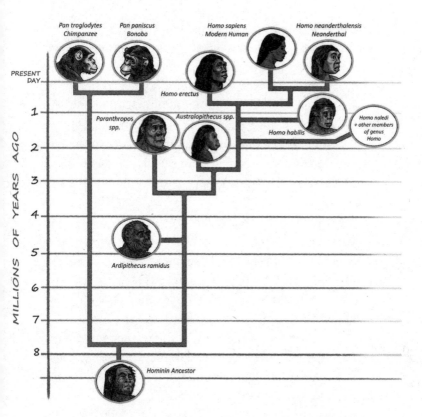

Figure 3 Evolutionary family tree ("phylogeny") showing relationships of living and extinct human species: chimps are our closest *living* relatives, but there were other species that were much closer to us in the past.

Life much more recently than the 400-million-year-old fish that first came onto land. For instance, the common ancestor between us and our closest living relatives, the chimpanzees, lived only about 6–8 million years ago (figure 3).[9] But chimps are only our closest *living* relatives; there were other, more closely related species that lived even more recently but that went extinct, some even less than 50,000 years ago.[10]

But what if there were other human species besides *Homo sapiens* living today—*Homo neanderthalensis*, *Australopithecus africanus*, and *Homo habilis*, for example? How would their presence shape our view of humanity? I'd like to think that it would open our eyes to our shared similarities and help us fill the gap between humans and other animals. But in reality, I'm guessing we would focus on our differences. Let's face it—if there were many more human species walking around today, sadly there'd also be many more flavors of discrimination.

Our species isn't exactly known for its tolerance, and maybe that's part of why so many people reject evolution—it puts us too close to the animals, it makes us an ape. Incidentally, all the other living apes besides us—chimps, gorillas, orangutans, and gibbons—are endangered because of us (whether through hunting, or habitat destruction). Maybe that same intolerance is why those other now-extinct human species disappeared about the same time *Homo sapiens* showed up.[11] But who knows, if we blurred the line between us and the rest of the animal kingdom, maybe we would see ourselves for what we are—just another recently evolved branch of the Tree of Life.

As an evolutionary biologist, I like to explain as much as I can through the gaze of history and ancestor-descendant relationships. As the saying goes, "Nothing in biology makes sense except in the light of evolution"—an apt but now hackneyed pronouncement of the eminent geneticist Theodosius Dobzhansky, one that evolutionary biologists have heard all too often. But no matter how clichéd Dobzhansky's words have become, they still ring true. You just can't explain life on Earth without seeing the connections between living things, connections we represent as the Tree of Life (figures 4A[12] and 4B). One living thing giving rise to another, passing along traits like those five fingers. Sometimes those five fingers become a bat's wing and sometimes a walrus's flipper. The origins are the same, evidence of our common ancestry. But evolution is also the things that don't make sense: our bad backs, our strange knee joints, nipples on males. You've got to take the bad with the good, and you can only explain it through the change from one living thing to another. Evolution doesn't make things perfect; it doesn't do anything to direct any individual or group toward some goal. If, as a living thing, you are fit and lucky enough to survive long enough to pass on your genes to your children, then you are a direct part of the evolutionary process. But even though evolution can be quite cruel (Darwin may have lost his religious faith over the excessive role of competition and death in natural selection[13]), it is not directed: it works without intent or vision toward the future. To quote another eminent biologist, François Jacob, "Evolution is a tinkerer, not an engineer."[14]

Incidentally, not reproducing doesn't mean you're not involved in the evolutionary process. Nonreproductive members of a population are indirectly involved in evolution: they aid it by helping reproductive members and their offspring survive. Consider, for example, the doting cousins with no children of their own, the sterile worker bees, and so on. They are all helping to pass along the genes that they share with individuals who *are* reproducing. In many cases they are passing along more of their own genes through that

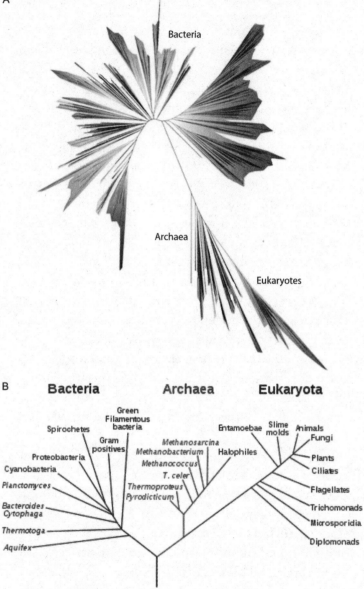

Figure 4 Two depictions of the Tree of Life. (A) Phylogeny with molecular evidence from Hug et al., "A New Tree of Life" (2016); (B) a simplified version showing other major divisions. (See plate 2.)

reproductively successful relative and the surviving offspring than if they had children themselves.

The products of the process of evolution are the survivors we see living today, modified bits and pieces that we can trace back from one ancestor to another. The evidence is in the genetic code (DNA and RNA) we all share, in the flesh and bones and behavior we can see and in fossils spanning nearly four billion years of change.

THE GREAT CHAIN OF BEING OR LADDER OF LIFE

Let's begin, as so many things do, with Aristotle. His view of life was a hierarchy called "The Ladder of Life" (or Scala Naturae in Latin). He set inanimate life at the bottom and humans at the top of the ladder (angels and gods and such would be added later by others as the ladder transformed into the "Great Chain of Being"; figure 5).[1] Aristotle's work was the most important influence on the West's understanding of the natural world for more than 1,000 years, and the Ladder of Life/Great Chain of Being metaphor is still how many people understand (or rather misunderstand) evolution, that is, as a linear process with bacteria and plants at the bottom as "primitive" and a straight line from fish → amphibians → reptiles → mammals and then humans as a distinct category at the top.

The problem with that antiquated, single-file view of how life is organized is that it makes you think of life as evolving in a straight line. Even people who accept evolution can get things wrong by taking this linear view. They see all other living things besides us as subordinate precursors leading to humans. They see chimpanzees as a step before humans, as if we evolved directly from them. With that view, they see

Figure 5 Representation of the Scala Naturae, with humans ("man") at the top of the hierarchy and lesser creatures below.

chimps and other apes as "primitive" antecedents of humans, which they are not. They also see a fish as a step before a frog, and a frog as a step before an alligator or some other reptile. Linear thinking can result not only in a very poor understanding of evolution, but also in a distorted sense of ourselves, especially when we push this thinking to include socially constructed views of human races and genders.

In his best-selling, overtly macho self-help book *12 Rules for Life*, Jordan B. Peterson argues in "Rule 1" that you should "Stand up straight with your shoulders back," building his argument based on the mistaken notion that our ancient shared ancestry with lobsters means we should somehow maintain certain shared innate behaviors with them.[2] Peterson explains how lobsters that are the victors of battles have upright postures: only losers slouch (he links posture to serotonin levels in both humans and crustaceans). Okay, I suppose slouching is bad, but we are not lobsters, nor did we evolve from them. We *do* share a common ancestor with them, however, we share common ancestors with all living things on Earth, and one of those ancestors gave us some common hormones like serotonin (which we have in common with many other multicellular life forms: pineapples have serotonin, for instance).[3] We did not receive our behavioral traits from some lobster ancestor, as implied by Peterson. And why cherry pick behaviors anyway?[4] Why not fixate on how some species of lobsters form "Conga lines," marching in long, single-file migrations?[5] Apparently, that part of lobster behavior isn't something he thought relevant to the modern-male human condition.

In the same book, Peterson argues that human sexuality and gender are essentially fixed because sex and distinct sexes were invented millions of years ago, and that the mother/father-child relationship is the oldest in our evolutionary history.[6] But, if you look closer at that evolutionary history or throughout the Tree of Life, asexual reproduction is certainly the oldest form of reproduction (many bacteria and archaea are clonal). There are also all-female species (like Amazon mollies, *Poecilia formosa*), multisex/intersex/no-sex species (as in many fungi that have "mating types" or slime molds that have hundreds of "sexes"), species that change sex (many fish species), and, indeed, there are species that reproduce without needing to find a mate (again, asexual reproduction) in every phylum of the animal kingdom. You could certainly argue for a variety of sexualities being

---

→

Figure 6A Evolution is the process that connects all life—all living things. The connections between living things on Earth are often depicted as a tree (see figure 4). But evolution can also be imagined as more of a web showing an unbroken, sprawling series of interconnected ancestor-to-descendant relationships (step 6). Importantly, extinction isn't depicted here (but see figure 6B). Starting from the first life—the first living thing, represented here by a single cell (with the white egg-shaped center)— expanding into more, new forms of unicellular life, represented by the other circles (with no holes in the center), each a different species (steps 10 to 9). Unicellular life keeps evolving, but so does multicellular life (step 8), represented by fungi, plants, and animals. If you focus on just the animal part of the web, you see a diversification of vertebrates, represented by fishes (step 7), and if you follow the vertebrate line, you can see that although fishes continue to evolve and diversify, one fish lineage expands and gives rise to the line that results in mammals including us humans (step 6'). And, as we evolved, so did the rest of life (full step 6), with more unicellular forms, too, all diversifying and evolving into the complete web of life (step 6). We humans tend to just focus on our line of descent (as in the new portion of 6'), but the entire evolutionary history of living things is much more complex. (See plate 3.)

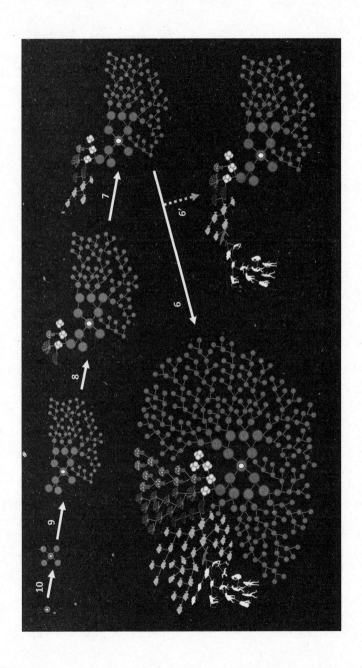

"age-old."[7] The misguided linear view of life—this false hierarchy—led some people to argue that white men are at the top of the evolutionary ladder, a convenient argument to enslave people of "lesser" races or to keep the womenfolk at home barefoot and pregnant. But this kind of thinking is insidious and, in its worst form, leads to racism, classism, sexism, transphobia, homophobia, and all manner of pernicious prejudices.[8]

To be fair, for a long time most Tree of Life illustrations (many by well-meaning people) almost always picked an image of an old white guy to represent all of humanity—"man"—on the branch that represents humans, even though a child, a women, or Jackie Chan could also have been picked to represent "man." Nothing against old white men, but representation matters; seeing a member of the

---

Figure 6B What is missing from figure 6A is extinction. Step 5 here shows that the earliest life forms gave rise to other life forms but went extinct along the way (the hole in the doughnut). There may be some fossil remains or other elements, but the center of this web of life is very old and has left hardly a trace except the descendants depicted as the younger outer ring, and a few fossil remains (the individual species inside the ring). The scattered life forms around us today (step 2) are just a fraction of all the life forms that have existed on Earth. We humans have tended to line up those life forms with what we see as the most advanced form, namely, ourselves, as the end of the line, with what we see as more "primitive" life forms following behind (step 1). Although the evolutionary line we picture is nothing but an artificial construct. Maybe it is all Aristotle's fault for getting us to think that way in the first place (see figure 5). Better to see the entire picture of evolution expanding in all directions instead of just focusing on the small part that led to us. (See plate 4.)

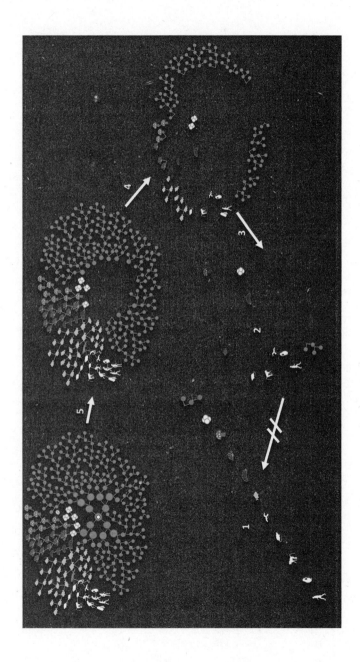

same subgroup of humans represented as the ideal human on nearly every evolutionary tree is damaging, and evolutionary biologists need to do better. We also tend to draw these trees with single-celled or "primitive" living things on the left, and humans on the right, which again gives the impression that evolution is linear; but because the descendants from the same node are "sister lineages" that can be flipped, you can also draw the tree with humans in the middle (like I did in figure 1B) for example, without any of the relationships depicted changing.

Let's go back to Aristotle's hierarchy. Most people today don't think of humans as having evolved from plants or fungi, so those who conceptualize evolution see that there was at least some distinct branching off from different forms of life. We often visualize the evolution of life as a branching tree, as in the Tree of Life, but there are other ways of visualizing it (figure 6A). Many people revert to the mistaken linear view because it is easy to organize the evolution of life with us humans at the end and everything else behind us (figure 6B, image 1). But it would be much more accurate if we visualized the evolution of life as a bush or bursting firecracker, expanding from a center. The living things around us are part of a process that started billions of years ago, with every living thing alive today nearly equidistant from a center point. Think of the origin of life as the "Big Birth," like the Big Bang that formed our universe, except with life radiating out from that center (first living thing, or last common ancestor of all living things), and with us and every other living species on the outer edge (figure 6B). The fossil forms from millions and billions of years ago are closer to that center; they are long extinct, as are many other forms of life from the past that have left no trace. Just like a fading firework in the sky, when the center has burst and faded into darkness, all that remains are the flaming edges expanding outward: these are the forms of life we see living today, and it is just a little snapshot of the evolutionary history of life (figure 6B, image 4).

Box 1
What Is Evolution?

Evolution is the grand process that led to the uninterrupted chains of ancestor-to-descendant relationships that connect all past and present life on Earth. Evolution is the changing of heritable features, with those changes leading to new kinds of life—new links on the chains. Evolution is the idea of common descent from a single origin—the first living thing—and the history of change that explains life on Earth. Evolution is all that wrapped up as a grand explanation or hypothesis; it is also a theory and a fact nested within other theories and facts. How we frame it doesn't matter that much, but we often hear evolution dismissed as "just a theory." Let's dive into why it can't be so easily tossed aside.

THEORIES OF EVOLUTION

I think about evolution all the time. I'll look at my hand and wondering about my five instead of eight fingers or how my eyes evolved to see that hand and beyond, or when looking at my twins and wondering about their identical genomes. Or when weeding the garden and getting a prick from a rose bush or feeling the sticky sap from some persistent weedy vines I thought I got rid of last week. I think about evolution when I mow my lawn: "Why do you grow back every week?" I snarl at the grass. "You know I'm just going to cut you again."

If our lawns did stop growing, we would have to explain that strange but natural phenomenon in scientific terms. Perhaps we could see if they only stopped growing if we used a certain type of mower, or because mowing the lawn changes the chemistry of the soil, thus preventing regrowth, or because the grass was sentient and somehow "learning." Some of these hypotheses are more plausible than others, some easier to test than others. Hypotheses are the possible answers to questions that can explain such phenomena.

For its part, in scientific terms, a theory is usually an over-arching, proven concept that has stood the test of time, such as the "theory of gravity." A theory is often the long standing "background knowledge" or "law" from which other theories and hypotheses are rooted. As used in everyday speech, however, "theory" is something that is questionable or that remains to be proven. But even in this everyday sense, evolution is *not* a theory.

There isn't a single theory of evolution—there are many theories. And more than one of those theories can be correct at the same time. One of those theories is "natural selection," made famous by Charles Darwin in his 1859 book *On the Origin of Species*,[1] which sparked "perhaps the greatest intellectual revolution experienced by mankind."[2]* Another is the theory of evolution by sexual selection, introduced by Darwin in the same book, a theory he expanded on in *The Descent of Man*.[3] There is also the "neutral theory of molecular evolution," introduced by Motoo Kimura.[4] These three theories are all valid and well-studied and complement, rather than contradict, one another.

There are other theories of evolution, too, that either have not held up to rigorous testing (because they don't fit new empirical evidence) and been retired, or that apply only in special circumstances (such as those which apply only to asexual organisms). People tend to equate "the theory of evolution" with "the theory of evolution by natural selection from Charles Darwin," but that's only one of the

---

* This quotation is from perhaps the twentieth century's most respected evolutionary biologists, Ernst Mayr, in his book *What Evolution Is*, which was published when he was 97 in 2001. (Mayr dedicated his book to Aristotle, and I remember wondering if they actually knew each other, so long and storied was Mayr's career.)

theories of evolution—and just one component of how evolution works. Darwin's theories of evolution are sometimes called "Darwinism," and many of the attacks on evolution focus on those nearly 200-year-old explanations, ignoring all other theories of evolution that have come about since.

So, what's the big deal about Darwin's theory of evolution by natural selection? Well, it is simple and intuitive, and it works. If the theory of natural selection were written as a motion by Mother Nature in a parliamentary meeting, it might read: "Whereas there is a great deal of variation among offspring, and whereas there is also a great deal of competition for resources, therefore be it resolved that only those organisms that are best able to survive in an environment may live on to reproduce" (figure 7). And, in shorthand, the gist of this motion might read: "Adults make babies, but only some of the babies survive to adulthood to make more babies." Or, simpler still, natural selection amounts to "survival of the fittest." For the record, Darwin used "survival of the fittest" as a shorthand explanation for natural selection only after another scientist, Herbert Spencer, coined the phrase; Darwin first used it in the fifth edition of *On the Origin of Species*, which is when it became associated with him.[5]

The term "fittest" as we modern evolutionary biologists use it, is related to reproductive fitness or success, not to how physically strong a living thing is. Many have misinterpreted natural selection as "survival of the strongest," with terrible consequences for how we treat one another and other living things on Earth. "Survival of the strongest" is a popular theme of the eugenics, white supremacy, and Nazi movements, which unfortunately are sometimes associated with Darwin as "social Darwinism,"[6] a notion that is neither

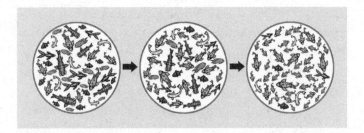

Figure 7   Natural Selection—Step 1: Make babies; Step 2: Babies struggle to survive because resources are limited, some die off; Step 3: Babies that prove to be best fit by surviving into adulthood make more babies like them. In this case the fish with spots are best fit and grow up to make more spotted fish. Note that all the variation is not lost, a change in the environment may favor fish with stripes next season. (See plate 5.)

Darwinian in origin nor relevant to how evolution actually works. Unfortunately, Darwin's own cousin Francis Galton was a eugenics leader (who even coined the term "eugenics"), as was Herbert Spencer (who, as noted above, coined the phrase "survival of the fittest"), and associations with these two men are partly why Darwin is linked with eugenic distortions of his ideas, even though he never made the same arguments that eugenicists did. On the other hand, Darwin should not be completely absolved from playing a role in the emergence of these consequential distortions, if only in an indirect role by not making it clear that his evolutionary theories did not apply to human societies.[7]

Natural selection alone can't explain all of evolution; even Darwin recognized that. That is why he came up with another explanation for the peacock's beautiful tail feathers and the bright coloration of some animals, which obviously make them more conspicuous to predators (hindering their survival). Hence the additional evolutionary theory of "sexual selection"—which explains that an organism may have traits that make it less likely to survive via natural selection because those same traits make the organism more attractive to its opposite sex and more likely to reproduce. You can see why sexual selection is sometimes seen as belonging under the umbrella of natural selection—and sometimes not.[8] To distinguish between them, think of sexual selection dealing with the competition for mates, whereas natural selection is all about survival. All these theories fit under Darwin's brief, but apt, definition of evolution as "descent with modification."[9]

We needed yet other theories of evolution to explain why living things vary: Why is there variation among different members of a species? Where does genetic variation come from? Why do you look different from your siblings even though you have the same parents? Why do identical twins behave differently despite having the same DNA? And for some of the theories that answer these and other similar questions, we needed to understand genetics—something Darwin and other early proponents of natural selection lacked the means to do. We also needed more theories, hypotheses, and conjectures to explain the many hereditary nuances of life and variation (including nongenetic environmental influences on development). We'll get into some of those nuances soon, but the point is that when someone says, "Evolution is just a theory," I say, "Nope, evolution is the scientific explanation for the origin and diversity of life. Evolution is a *fact*. Descent with modification is a *fact*." We can't explain the world and the diversity of living things without evolution. And we can't disprove evolution with what we observe in the natural world.

One of the best lines of evidence for evolution is that we all share the building blocks of life, DNA. We can use DNA to put together the Tree of Life and show the connections between every living thing on Earth, including ourselves.* Just because we are the only living things that can put together the Tree of Life doesn't mean we stand outside it. We are still a part of the Tree of Life, and, as on a map, we can use it to see where we are and where we came from.

In the end, let's remember that natural selection was not the first theory of evolution, not the only theory, and not the last theory of evolution. Despite the existence of other theories of evolution, natural selection is an important insight into the mechanisms of how living things evolve, perhaps the most important insight. In many ways that theory evolved much like the mechanisms it attempts to explain, it survived over competing theories and continues to be shaped over time. Let's look more closely at its origins.

---

* Of course, we should also use the physical form and structure of living things (their "morphological features") in putting together the Tree of Life, so that we can include other species, both existing and extinct (using fossils), to aid and complement the DNA-based evidence.

FROM SO SIMPLE A BEGINNING

Illustrated by Ethan Kocak

There isn't a good movie about Charles Darwin's life, which is a shame, so I asked famed illustrator Ethan Kocak to sketch out a plot for a Darwin movie that I wrote. The title was Ethan's idea, and it, along with the last line, are based on quotes from *On the Origin of Species*.

Scene 1 We see a distraught Captain Pringle Stokes, first captain of the *Beagle*, taking the ship on its maiden voyage off of Tierra del Fuego and taking his own life.

Scene 2 and 3 (top)   Young Mr. Darwin, avid collector of beetles, spies another beetle while holding one in each hand. When he pops one into his mouth for safe keeping, it releases an irritant that causes his mouth to foam. He loses all three specimens. Back home, his father is upset with

YOU CARE FOR NOTHING BUT SHOOTING, DOGS, AND RAT-CATCHING, AND YOU WILL BE A DISGRACE TO YOURSELF AND ALL YOUR FAMILY!

I WANT TO SEE THE WORLD! I PASSED OUT IN MEDICAL SCHOOL AND WAS BORED BY DIVINITY SCHOOL!

I WANT TO CONTINUE GRANDFATHER ERASMUS'S LEGACY, NOT YOURS.

his aimless son (that's a real quote from Dad). Darwin doesn't want to be a dull physician like his father, but more like his grandfather Erasmus, who was a physician but also a philosopher of nature who wrote weirdly erotic plant poetry that presaged natural selection.

Scene 4 Darwin leaves home on HMS *Beagle* on what was supposed to be a two-year voyage but that actually lasted almost five. He was often seasick, but when they did make landfall on their Southern Hemisphere circumnavigation (focusing on South America), he uncovered fossils of beasts small and large, witnessed earthquakes and discovered new species; he had a complex relationship with Captain FitzRoy, who was in search of evidence of biblical events. What Darwin was finding instead was evidence of a much older planet than what FitzRoy thought the Bible suggested. Ironically, it was FitzRoy who gave Darwin a copy of Lyell's *Principles of Geology*, which was one of the first texts to argue for a very old Earth.

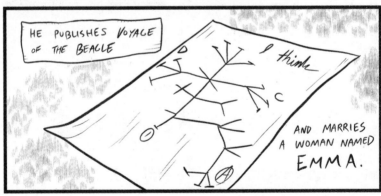

Scene 5 Darwin returns to England and becomes a famed naturalist based on his *Beagle* adventures. He enters high scientific society, and his specimens are being analyzed by mentors like ornithologist John Gould, who chastised him for his poor field cataloguing of the Galapagos finches. But Darwin intuits that these finches (and other Galapagos animals) had diversified on these islands from South American ancestors. These thoughts lead him to thinking about evolution, and he begins writing the seeds of what will become *On the Origin of Species*. In 1839, he publishes the first version of what we now refer to as his *Voyage of the Beagle* and marries his cousin Emma Wedgwood, who is deeply religious.

SHE WRITES HIM ABOUT HER WORRIES ABOUT HER DEEP FAITH AND HIS LACK OF IT.

HE CONTINUED TO WORK FROM HIS HOME

DESPITE HIS HEARTBREAK OVER LOSING CHILDREN AND WONDERING IF IT IS BECAUSE HE AND EMMA ARE FIRST COUSINS.

HE GETS A LETTER FROM ALFRED RUSSEL WALLACE.

NATURAL SELECTION IS A THING.

-Alfie

DARWIN RECALLS HIS MENTOR ROBERT GRANT, WHO STOLE HIS WORK & PUBLISHED IT...

Scene 6 Darwin works diligently at home on barnacles, plants, birds, and so on—his wife would give birth to ten children, but when several of them die, Darwin worries if their deaths are due to his wife and he being first cousins. (Paul Bettany plays a feeble Darwin in a movie called *Creation* that does a good job of covering this troubling consanguineous aspect of Darwin's life.) At his lowest point, he gets a letter from another naturalist, Alfred Russel Wallace, who is doing fieldwork in Malaya. The letter includes a draft manuscript that is very similar to the ideas about natural selection that Darwin had been secretly working on for years.

DARWIN SPILLS THE BEANS TO LYELL AND HOOKER.

THEY PRESENT HIS WORK ALONG WITH WALLACE'S TO THE LINNEAN SOCIETY.

MEANWHILE...

DARWIN LOSES HIS SON.

AND WALLACE IS STILL IN THE MALAY ARCHIPELAGO

HE FINISHES WORK ON "ORIGIN of SPECIES" AND THERE'S QUICKLY GREAT DEBATES ON THE STREETS.

WRONG!

Scene 7 After Darwin shares Wallace's letter with his scientific colleagues, they announce the theory of evolution by natural selection to the world with Darwin and Wallace's work presented jointly. At this time, Darwin is burying his youngest son, Charles, and Wallace is still in the field, completely unaware of all that is going on (he'd be pretty cool with it though).

Scene 8 (previous page) Darwin finally completes his magnum opus, *On the Origin of Species*, and it causes a huge stir, including the great debate between Thomas Henry Huxley and Bishop Samuel "Soapy Sam" Wilberforce.

Scene 9 Darwin's old traveling companion from the *Beagle*, Captain FitzRoy, is at the debate muttering about the Bible; he would later follow the same sad path as the first captain of that great ship.

Scene 10 Darwin would die many years later (1882), at the age of seventy-three, largely vindicated—he is DarWIN after all, not DarLOSE (to quote a famous Dana Carvey comedy sketch).

Although we dramatized some events here (hey, it's a cartoon movie; FitzRoy had a much bloodier death than we depicted and the details of the Huxley-Wilberforce debate are controversial), it is largely based on the realities of Darwin's life as told by several biographies, including his own.[1]

ONE OF DARWIN'S MOST IMPORTANT EARLY INFLUENCES WAS FROM A FORMER ENSLAVED MAN NAMED JOHN EDMONSTONE – WHO TAUGHT TAXIDERMY AND TOLD DARWIN OF HIS WONDERFUL TROPICAL ADVENTURES AS A NATURALIST. PERHAPS WITHOUT EDMONSTONE DARWIN NEVER GOES ON THE BEAGLE AND WE DON'T GET "THE ORIGIN OF SPECIES"

Postscript "From So Simple a Beginning." Drawn by Ethan Kocak.

I wish we knew more about Darwin's early mentor, John Edmonstone. Given Edmonstone's background and experience and Darwin's inquisitiveness, this seems to have been an important relationship in Darwin's life.[2] Edmonstone had been enslaved in South America working for a naturalist; one can envision Darwin being enchanted by stories of collecting and travel as he learned taxidermy from Edmonstone. Perhaps these tales helped him make the decision to join the crew of HMS *Beagle* years later.

THE SELECTION OF NATURAL SELECTION

Perhaps the most controversial argument in *On the Origin of Species* was not natural selection but that there is a single origin of life—a common ancestor of all living things—on Earth. As Darwin put it: "All the organic beings which have ever lived on this Earth have descended from some one primordial form."[1] That was a huge leap forward in thinking; it was particularly powerful when viewed with natural selection as the causative mechanism to explain how we got the diversity of living things from that single common ancestor. (We will discuss later if the first living thing was also the most recent common ancestor of all living things today. Perhaps it was not.) Darwin was saying that every living thing on Earth was related. At the time, many people, including popular scientists like Louis Agassiz, didn't even think all human races had a single evolutionary origin (most importantly, and rare for the time, Darwin thought they had and that was part of his abolitionist stance), and earlier theories of evolution assumed there were many origins of life throughout time.[2]

The first major scientific theory of evolution was proposed by Jean-Baptiste Lamarck (1744–1829), often explained as the "accumulation of acquired traits."[3] Lamarck thought that different species had independent origins and that they went from simple to complex as they picked up features through life, which they passed on to their offspring. The Lamarckian example everyone cites is a giraffe stretching its neck, which would cause it to grow longer; and so the offspring of that giraffe would have a longer neck than its parent as a result. We know through many experiments that species don't pass down traits that they acquire in their lifetimes. For instance, in one experiment, a scientist cut the tails off generation after generation of lab rats, and each generation of rats had normal tails when they were born, whereas, under Lamarck's theory, they should have had shortened or no tails at all.[4] But we know intuitively now that acquired traits aren't passed down from one generation to the next—you don't have blind children because you lost your eyesight; you don't pass down your muscular physique because you work out at the gym twice a week; and the children of Michael Phelps don't have webbed feet.

I saw an antivaccine ("antivax") social media post once about how a couple who wanted kids also wanted to get the flu and measles so their future children would gain immunity—that's Lamarckian, and wrong. You don't pass down your immune response to your children in this way; otherwise, the flu would have died off with the first generations to get it. The antivax version of herd immunity, where folks don't take vaccines and fight off diseases "naturally" on their own, is actually Darwinian because it requires a lot more people (all those who can't fight off the diseases on their own) to die, leaving behind those with a natural immunity. This antivax herd immunity is very different from the real herd immunity advocated by modern scientists, in which most of us would have vaccine-aided immunity against a given disease and avoid getting that disease in a deadly form. The scientific version of herd immunity would stop the spread of disease with far fewer deaths than required by the antivaccination folks[5]—in fact, at least 14 million people are estimated to have been saved by COVID-19 vaccinations as of 2022.[6] Good thing most of us have moved on from Lamarck, and good thing Darwin did too.

Besides Lamarck, another influence on Darwin was an initially anonymous and popular book, titled *Vestiges of the Natural History of Creation*.[7] That book promoted divine intervention as transforming one form into another (including inanimate objects like rocks and stars). Although Darwin disagreed with that idea, Darwin recognized that *Vestiges* did improve upon Lamarck and lay the groundwork for his theory at least in terms of priming the public for a new single-origin-of-life theory about "evolution" (a term that did not exist in the modern sense until later). Scholars in much of Darwin's time used the term "transmutation" instead of "evolution," although Darwin did use "evolved" before many others did; indeed, "evolved" is literally the last word of the first edition of *On the Origin of Species*.

Darwin would probably have never published or completed his so-called abstract on evolution, *On the Origin of Species*, if he hadn't received a letter from a fellow biologist, Alfred Russel Wallace that left him profoundly upset. In what is now Indonesia, Wallace was working on his own masterpiece on animal distributions,[8] when a great insight struck him between bouts of malarial fever dreams.[9] He hurriedly wrote Darwin a letter and sent along a draft manuscript about his own ideas on how species evolved, and those

ideas were nearly the same as Darwin's. Having been thinking, experimenting, amassing evidence, and procrastinat-
ing on the problem for twenty years, Darwin was stunned.[10]
Wallace and Darwin both recognized that species come from
other species through a persistent struggle for survival (see
figure 7). Darwin had understood this for years, but had
held back from publishing because he wanted more evidence. In fact, he wanted to compile a huge, encyclopedic,
multivolume text that he probably would have preferred to
have published posthumously. But this was his grand idea,
a single origin of all life—of all living things—and the struggle for survival that leads to adapting and evolving species:
he couldn't let the upstart Wallace be the first to publish it
and claim all the credit. Darwin's prominent science friends,
Joseph Hooker and Charles Lyell, presented Wallace's paper
along with Darwin's long-concealed work to the scientific
elite and published it in 1858[11] (all unbeknownst to Wallace still in Malaya, and with the tacit blessing of Darwin
who was burying his recently deceased toddler Charles at
the same time).

The next year, 1859, Darwin published what would be an
incomplete, somewhat hastily put together masterpiece, *On
the Origin of Species by Means of Natural Selection, or the Preservation of Favoured Races in the Struggle for Life* (you can see why
I just use the shorthand *On the Origin of Species*).[12] It would be
the most consequential and readable nonfiction book ever
written in the West[13]* (sorry folks who have pretended to

---

* Unfortunately, we don't have much information about non-
Western thought on the evolution of life, although it would not be
surprising if natural selection had independently been discovered in

read Newton's *Principia*) and one that would see six editions. Arguably, despite its flaws, the first edition is the best one because it is the humblest and most straightforward version: Darwin wrote apologetically for not having more information about the sources of variation and the lack of fossil evidence at the time. Darwin knew nothing about genetics (we wouldn't know about the structure of DNA for almost 100 years after the first edition was published), and he went on to speculate (perhaps too much) about the origins of heritability in his future work.[14] Unfortunately, Darwin would die without ever knowing about the contemporary and pioneering work in the study of heredity by a humble Austrian monk named Gregor Mendel.

China, India, or western Africa. Read S. Frederick Starr's *Lost Enlightenment*, for a scholarly take on some discoveries by Central Asian scholars on space, time, religion, medicine, and the natural world that predate or parallel those of the West. The first-century Iraqi scholar Al-Jahiz also expressed pre-Darwinian evolutionary ideas in his *Book of Animals*. Rebecca Scott, in *Darwin's Ghosts*, does an excellent job explaining how the early evolutionary writings of others may have influenced Darwin. The contributions to early natural history from non-Western scholars merit more lengthy discussion than I can provide here. To quote from Darwin's 1858 joint paper with Wallace introducing natural selection: "This sketch is *most* imperfect; but in so short a space I cannot make it better. Your imagination must fill up very wide blanks."

MENDEL AND THE
MAINTENANCE OF VARIATION

Many biologists would quickly come to accept Darwin's explanation that all living things evolved from a single ancestral form. "Yes," they would agree, "evolution is a thing." But they had trouble accepting the mechanisms of descent proposed by Darwin. The going hypothesis at the time was that offspring were a mix of their parents, the result of what was called "blending inheritance."[1] Through blending inheritance, a red mother and white father would produce offspring with a "blend" of those traits, namely, pink daughters and sons. If blending inheritance took place, that would mean variation was lost with each generation (i.e., two different-looking parents would produce intermediate offspring).[2] Enter Gregor Mendel, an Austrian monk with a love for science and peas.

In one of science's great insights, Mendel discovered how variation was maintained over generations.[3] He mixed hundreds of pure strains of pea plants with different traits (like purple flowers and white flowers) and noticed that he got all purple flowers in the first generation (variation is lost, but not blended). More importantly, in the second generation, when he mated all those first-generation purple-flowered plants together, the resulting offspring were about three quarters purple flowered and one-quarter white (the trait that was lost is back). From this result and additional experiments on other traits, Mendel deduced that the controls for those traits must be passed down to the offspring in pairs. Imagine those controls coming in the form of one dominant version (let's call it "B" from the female) and one recessive version (call it "b" from the male). In the first generation, as the different versions combine in the offspring, all the dominant copies block all recessive copies from being expressed: traits B and b are paired up in the offspring as Bb and you see only one phenotype—the dominant B copy expressing purple flowers. But, in the second generation, when the Bb mothers and Bb fathers meet up and the copies of the genes are split up into dominant B and recessive b in their offspring, the two recessive b traits can appear together (as bb—recessive copy b without dominant copy B) in a quarter of the offspring (figure 8), resulting in the reappearance of that recessive phenotypic trait (bb, in this case, white flowers). Mendel also noticed that all the traits he was studying (seed color, plant height, and so on) had the same ratio (¾ dominant, ¼ recessive) in the offspring of the second generation after cross-breeding pure strains, but that those traits were not linked to each other (e.g., purple flowers and tall plants didn't always appear together). He realized that those traits were somehow being mixed and divided as they were passed down from generation to generation (in what we now call "recombination"). For me, Mendel's discovery of genetic inheritance patterns—now known as the "law of segregation" (each trait is represented by a pair of gene copies or "alleles") and the "law of independent assortment" (each

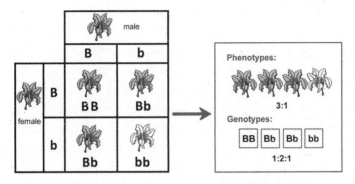

Figure 8 Crossing dark-flowered pea plants yielded three-quarters dark-flowered plants and one-quarter white. Why? Mendel deduced there were pairs of what we now know as "alleles" (different versions of genes that control traits like flower color) that separated during reproduction (the dominant B and the recessive b). Whenever the dominant B allele was present, a dark flower appeared, and whenever a recessive b was present without the dominant B in the combined seed, a white flower appeared. What Mendel's experiment showed us is how to separate the genotype from the phenotype to explain the maintenance of variation. (See plate 6.)

trait is sorted independently) is one of the most remarkably perceptive scientific insights in history. Think of it, Mendel figured out how chromosomes and genes maintain variation, without actually knowing about either and without having a microscope or DNA sequencer, but simply by observing what happens when you grow and cross-pollinate pea plants.[4]* That's right, a monk growing peas figured out

---

* Although Mendel's understanding of what he discovered might be a little overstated, the ground-breaking genetic insights of this Austrian monk have stood the test of time (2022 marked the 200th anniversary of his birth). I'd put his insights discovered in obscurity up there with Einstein conceiving relativity while working as a lowly patent clerk.

the fundamentals of genetics! Now, not all traits are so easy to tease apart (not everyone has either a big nose or a small nose) because some are "continuous" (they are found in a range, like nose sizes, rather than being "discrete"). What Mendel did was help establish that there are genetic controls (in the genotype) to maintaining physical variation (in the phenotype) and to prove that "blending inheritance" was not a fatal flaw for explaining evolution. Mendel's work rescued Darwin's.

How important are the mechanisms for producing and maintaining variation that Mendel discovered? Well, it explains sex and why we traded immortality for it. Only sexually reproducing organisms follow these Mendelian laws of inheritance. On the other hand, asexually reproducing organisms, like many microbes, fungi, some plants, but also some vertebrates, do not: they can just clone themselves, copying their DNA in its entirety with each generation, which allows them to pass down all their genes to their clones (hence immortality). Their genomes are immortal as long as those clones clone on. But this clonality also makes them more vulnerable to changing environmental conditions because, lacking genetic variation, they are less able to adapt to change.[5]

If only Darwin had read Mendel's work, he might have understood how genetic variation was maintained and his theory of evolution by natural selection would have been even stronger.[6] Both Darwin and Mendel would die before the world noticed the simple but magnificent genetic work of this unassuming monk. It wasn't until the start of the twentieth century that people understood that genetics was the link to solving the mystery of inheritance.[7] Others would discover that the two copies of each trait Mendel noted were the different alleles on sister pairs of chromosomes and that independent traits were on different places on the chromosome (or on another set), which is why they segregated separately. It was because of the work of Mendel and his successors that we learned that there was a dominant version of a gene and an often-masked recessive version—and how that explained how a recessive trait could "reappear" after several generations of being lost, as in certain diseases like hemophilia. It also explained why blood types appeared in predictable "Mendelian ratios."[8] Although Mendel's discoveries were a good start, it would not be until the middle of the following century that we found the real source of genetic variation in our DNA. And, as Mendel himself discovered, when working on other organisms, not every organism is as simple as a pea plant.[9]*

---

* The explanation of Mendelian genetics I use here can be found in nearly every introductory biology textbook (see endnote 9 to this chapter). Mendel was lucky to work on pea plants with "discrete" features (like flower color) that segregated nicely, he would later try studying bees. Unlike Mendel's easy-peasy pea plants, bees have a much more complicated pattern of heritable traits (read up on quantitative genetics of continuous traits or of the equally amazing discovery and use of white-eyed fruit flies in modern genetics), which is why Mendel, as the proto-geneticist beekeeper, made no discoveries about inheritance in bees. (That must have really stung.) Like many underappreciated academics, he went on to a life of obscurity as an administrator.

MUTANTS AND MUTATIONS

Maintaining variation is one thing, but what Mendel and Darwin could not know is that you can't have new variation without mistakes. Mistakes in our DNA code are called "mutations," and these are the roots of genetic variation. For many people, the word "mutation" conjures up the image of malformed creatures. But, in fact, we are all mutants: mutations are just natural variation in DNA. They happen in the copying of DNA: they are happening right now, all the time, as our cells divide. Just as even the best proofreaders at the *New Yorker* occasionally let "dessert" pass for "desert," so, too, does the DNA replication machine sometimes write "ACTTG" as "ACTG." And just as a proofreader's oversight error can let a word meaning "a delicious, sweet dish" turn into a word meaning "a barren landscape," the coding error of the DNA replication machine can turn a normal body into a malformed one. But out of three billion nucleotide base pairs in the sex cells (eggs and sperm) representing a single human genome,[1] pairs made up of molecules of the four nucleotide bases, adenine, cytosine, thymine, and guanine—A's, C's, T's, and G's—we only get about 10 to 100 or so copying errors in the DNA of a child that are not inherited from the child's parents.[2] A proofreading rate of about 10^{-8} (or 0.00000001 percent) that any magazine editor would be envious of.

DNA carries the genetic code that is the blueprint for the proteins that make us and control and run most of what we do. An error in the code usually does . . . absolutely nothing. One of the most surprising discoveries in our understanding of mutations and evolution was described in 1968 by Motoo Kimura, who found that many genetic mutations do nothing to change the abilities of an organism either to survive or to reproduce, so that the products of the DNA code that include how we look—our "external phenotype"—remain unchanged.[3] Kimura called his explanation of this phenomenon the "neutral theory of molecular evolution,"[4] and it is now closely associated with "genetic drift," as one of the ways species diverge, through slow, gradual, "drifting" nonadaptive change in DNA sequences, and a complement to natural selection.

Remember, evolutionary change is heritable change. Not every mutation has the same probability of happening, nor is every mutation passed down. Neither is every mutation a so-called "point" mutation, which is a change of one DNA nucleotide to another (e.g., an A to a C). The "ACTTG" → "ACTG" example above is called a "deletion," where a T nucleotide is deleted. There are also "insertions," where an extra nucleotide is added. These "indels" (short for "insertion/deletions') can have profound consequences depending on where that mutation happens. An indel can change how the DNA code is read, so that the protein can no longer be produced, and not making a certain kind of protein can have terrible consequences for the phenotype.

Despite what you may have learned, evolution is *not* random. (At least no more random than death and reproduction.) The mutation examples described above can appear spontaneously, but some parts of the genome are more susceptible to change/error than others. But, in much of the genome, it doesn't matter what changes happen because they take place in noncoding parts outside of the reading frames of genes. Other parts of the genome undergo "purifying selection," and any mutations are quickly weeded out. Mutations may happen spontaneously and accidentally as our cells divide, but they do not become part of the DNA code that is passed down to future generations, unless they make it into the DNA of the sperm or egg cells that combine to make those next generations. These spontaneous mutations become part of the evolutionary history of a species only when they are spread to a representative number of individuals in a population. How do they spread? As individual organisms reproduce, they pass along their DNA code (including mutations) to their offspring, and these changes are ultimately spread (or lost) through genetic drift (when neutral) or natural selection (when adaptive). Under favorable conditions, or when the mutations are beneficial, they can spread very quickly—or they can disappear in an instant.

Also, most mutations won't kill you, despite what you may have heard. But you might be wondering about the ones that will. (Everyone loves a good deadly mutation story.) If a mutation leads to death, why does that mutation persist? There are several reasons why it could. For example, some mutations, such as those which lead to Huntington's disease,[5] are expressed later in life, after parents have already passed their genes down to the next generation. Our mutation load grows with age, as our cells stop replicating as efficiently as they did in our youth (part of the natural stages of getting old called "senescence")—aided by smoking, drinking, or eating too many sugary snacks and doing so many of the things that make life worth living. But, if those later-expressed mutations aren't passed on in the parents' sperm and eggs to the next generation, they have nothing to do with evolution. They happen in our normal body ("somatic") cells, not in our sex cells ("gametes"). Notably, because every woman is born with all the eggs she will ever have, her own eggs developed with her inside of her mother; so you were an egg inside of your mother as she was developing inside of your grandmother (so you might want to ask grandma what she was eating, drinking, and smoking back them).

Sometimes, in what is called "genetic hitchhiking," a deadly mutation can be passed down by being located on a chromosome near a beneficial or required gene sequence. Although Mendel was correct about genes sorting independently, it depends on where they sit on a chromosome: genes that are physically close together on a chromosome are more likely to be inherited together generation after generation.[6] Think of it like loosely shuffling a deck of cards—if you shuffle only a few times, most of the cards that started next to one another will *stay* near one another. That shuffling takes place between matching ("homologous") chromosome pairs that are then separated. After shuffling, roughly half your DNA (as one of the two sets of chromosomes) is rearranged and set aside for making the sperm or egg. We make a lot of these gametes, and each will hold a unique combination of genes represented by a half set of our chromosomes. This process called "recombination" is responsible for making the unique mix of your parents' DNA that is you (when their egg and sperm came together). Think of your parents' genomes as two decks of cards (one is your mother's and one your father's): the decks are each shuffled independently and half of each deck is taken to make a new full deck which represents you (figure 9). Think of most mutations as the many small nicks and marks the cards in a deck may collect over time, most of which have no consequences to any game you play. On the other hand, if you lose a card while shuffling, that will be a much bigger problem when the lost card turns up missing from either of the two half decks (your mother's egg or your father's sperm) that make up the new full deck of cards (you).*

Recombination is responsible for removing mutations (even adaptive ones), as the beneficial mutations sometimes get shuffled out just as easily as the harmful ones. In that way, recombination is a force for both creating and removing genetic variation.

It is important to remember that even though most genetic mutations do not have major consequences for evolution, small mutations can still cause big differences in the phenotype. For instance, a mutation that changes a single nucleotide base in a specific DNA position from an A to a G causes insufficient formation of bone from cartilage, resulting in a form of dwarfism.[7] An example of the near opposite case is Joseph Merrick, also known as the "Elephant Man," who had Proteus syndrome, which results from a mutation in the

---

* If you don't like cards, you may prefer the salad metaphor that I use later to explain a similar consequence of recombination.

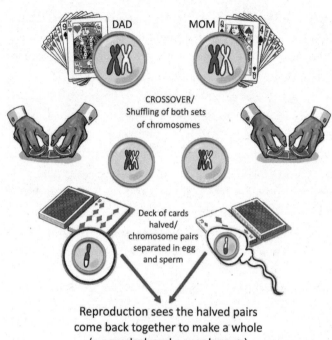

DAD   MOM

CROSSOVER/
Shuffling of both sets
of chromosomes

Deck of cards
halved/
chromosome pairs
separated in egg
and sperm

Reproduction sees the halved pairs
come back together to make a whole
(a new deck and a new human)

AKT1 gene that can appear in just one cell before spreading like a cancer and causing an abnormal overgrowth of bones.[8] Some of these mutations that can change the phenotype are less significant than others. One of my favorite less significant mutations involves a single nucleotide change in the DNA coding region for your olfactory receptors, this mutation causes loss of the ability to smell the sulfury stink of aspartic acid. This loss gives those who have the mutation the impression that their urine doesn't smell bad after eating asparagus, when, of course, it really does, they just can't smell it.[9] There is even a similar single-nucleotide mutation that causes cilantro to taste like soap to some people.[10]

Now, I have a confession: I have a rare, potentially consequential mutation. While trying to make babies, my wife and I ran into fertility problems, as many couples do. My wife's

---

Figure 9 "Deck of cards" metaphor for recombination. Your parents each have twenty-three pairs of chromosomes (forty-six total). Each egg from your mother and each sperm cell from your father has half the set (twenty-three pairs total). During meiosis the parental set is "shuffled" through genetic crossover, where pieces from the two different sets are recombined. During this recombination, the deck is split into half, so that only one half of the recombined chromosomes is in each sperm cell and egg. An egg and sperm cell come together during reproduction, bringing the total number of chromosomes back to forty-six. As recombination takes place, some mutations will be lost (whether beneficial, harmful, or neither), but this process also produces the unique combination of features that make you and your siblings different from one another and from your parents (in addition to the unique mutations that you gained in this process). (Please note that the process of meiosis from which sperm cells and eggs are made is far more complicated than summarized here.)

miscarriage in the eighth week of her pregnancy shocked and saddened us, but, looking ahead, it also concerned us, so we had the miscarried fetus tested. It turned out to be a "triploid," with three sets of chromosomes (sixty-nine total) instead of the usual two (forty-six total). One extra chromosome can lead to conditions like Down syndrome (forty-seven total chromosomes), but a complete extra set of chromosomes is almost always fatal before birth. To understand why it happened, my wife and I had our own DNA tested. It turns out I have an "inversion" on chromosome 8: part of one arm of that chromosome is on backwards. "That's why I'm like this," I deadpan to my students to see if they get it (as far as I know I show no external signs of the inversion). Did that inversion cause our miscarriage? Probably not, but the discovery did lead to genetic testing of the fertilized egg that would eventually become our identical twins (more on them later).

Significant changes and movements of pieces of chromosomes, like my inversion, were once thought to be rare. Then in the 1940s, Barbara McClintock discovered "jumping genes" (really the movement of sections of DNA from one part of a genome to another) that could lead to phenotypic changes, including variation in pigmentation patterns or even increased susceptibility to certain diseases like colon cancer—her research led to an entirely new avenue for understanding hereditable change.[11] McClintock helped us understand that these changes in our DNA can have either profound effects or no effects at all, depending on where they happen or what they lead to. Since then, we have also learned that genetic variation can come not just from within an organism, but also from without.

In the 1960s, Lynn Margulis popularized the theory that sometimes it is cooperation and not competition between organisms that can lead to major changes in evolution. Margulis described how the incorporation of free-living cells like mitochondria and chloroplasts into other cells led to a "symbiosis" that formed entirely new types of organisms.[12] These new organisms eventually would include what became plants, fungi, and animals. By making symbiotic relationships with chloroplasts and mitochondria—the energy-producing "powerhouses" of their cells–the new organisms were able to outcompete other organisms.[13] There might even be genes that jump around the Tree of Life via viruses and lead to convergent phenotypes in different organisms, which is one hypothesis about how different fish, mammals, and some other organisms evolved placenta-like organs independently.[14] Perhaps we will discover that other major organ systems also evolved convergently with the aid of viruses. (I wouldn't be surprised to learn that eyes or brains in vertebrates and various distantly related invertebrate lineages, independently evolved in this way.)[15]

Surprisingly, more of our human genome originated from viruses than is "functional" (i.e., that actually code for proteins). At least 9 percent of our genome has viral origins, while only about 1.5 percent is protein coding.[16] Even stranger, most of those functional genes we have can also be found in the genomes of invertebrates, or even those of plants and fungi.[17] I'm sure if Lynn Margulis (who died in 2011) had been an active scientist in the genomic era, she would have wanted us to revisit what it means to be human, or even an "individual." She argued for a "holobiont" view where a single organism is the individual + its symbionts (like the bacteria we carry in our guts and on our skin). She even argued that the interactions among symbionts was responsible for the formation of new species through a process she called "symbiogenesis."[18] These ideas never really took hold, but read the fascinating book *I Contain Multitudes* by Ed Yong for more on that debate,[19] including discussions of the works of great modern symbiosis researchers like Nancy Moran[20] and Margaret McFall-Ngai,[21] both of whom provide a nuanced and evidence-based take on symbiosis and evolution.

Our understanding of genetic variation has certainly advanced a great deal since Mendel's time. Although Mendel had the good fortune to study genes in pea plants that expressed discrete traits like "tall" versus "short" and "purple" versus "white," there isn't usually a single gene for each discrete trait like the features Mendel found in his pea plants. As shown in the examples above, the genetics of variation is much more complicated than either Darwin or Mendel could have ever imagined. But science, much like evolution through natural selection, moves incrementally, and often "one funeral at a time."[22]* Although, luckily, we have evidence for most of the steps in the evolution of scientific ideas preserved in journal articles and books, that's not the case for organic evolution.

---

* This quote is the often-paraphrased version of the German physicist Max Planck's statement in his 1950 autobiography: "A new scientific truth does not triumph by convincing its opponents and making them see the light, but rather because its opponents eventually die, and a new generation grows up that is familiar with it."

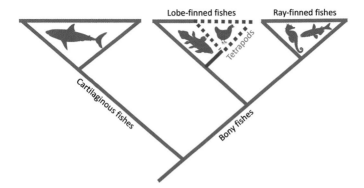

Plate 1 The Vertebrate Tree of Life is the same as the Fish Tree of Life (since we can call all vertebrates "fish" because of our shared ancestry); four-limbed vertebrates ("tetrapods") are in the red and dashed area of the tree (represented by a hen, but we humans belong in there, too—along with other mammals and with birds, reptiles, and amphibians) as members of the bony fish lineage; specifically, we are lobe-finned fish (in which lineage lungfishes and coelacanths also belong). We could call everything in this tree a "fish" because they all have gills and fins at some stage, even though the animals in the dashed portion of the tree are mainly found on land and have modified gills and fins, as we do.

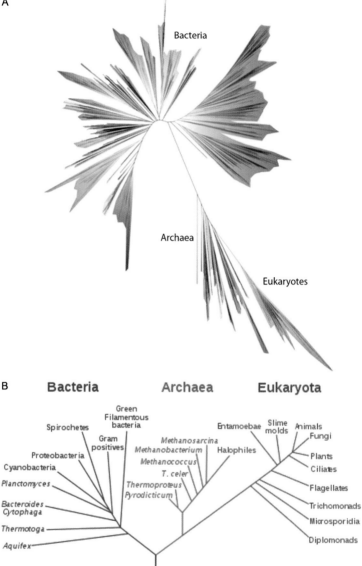

Plate 2 Two depictions of the Tree of Life. (A) phylogeny with molecular evidence from Hug et al., "A New Tree of Life" (2016); (B) a simplified version showing other major divisions.

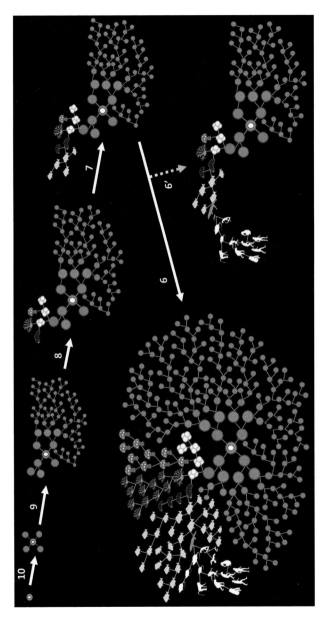

Plate 3

Figure caption on next page

Plate 3 Evolution is the process that connects all life—all living things. The connections between living things on Earth are often depicted as a tree (see plate 2). But evolution can also be imagined as more of a web showing an unbroken, sprawling series of interconnected ancestor-to-descendant relationships (step 6). Importantly, extinction isn't depicted here (but see plate 4). Starting from the first life—the first living thing, represented here by a single cell (with the white egg-shaped center)—expanding into more, new forms of unicellular life, represented by the other circles (with no holes in the center), each a different species (steps 10 to 9). Unicellular life keeps evolving, but so does multicellular life (step 8), represented by fungi, plants, and animals. If you focus on just the animal part of the web, you see a diversification of vertebrates, represented by fishes (step 7), and if you follow the vertebrate line, you can see that although fishes continue to evolve and diversify, one fish lineage expands and gives rise to the line that results in mammals including us humans (step 6'). And, as we evolved, so did the rest of life (full step 6), with more unicellular forms, too, all diversifying and evolving into the complete web of life (step 6). We humans tend to just focus on our line of descent (as in the new portion of 6'), but the entire evolutionary history of living things is much more complex.

Plate 4 What is missing from plate 3 is extinction. Step 5 here shows that the earliest life forms gave rise to other life forms but went extinct along the way (the hole in the doughnut). There may be some fossil remains or other elements, but the center of this web of life is very old and has left hardly a trace except the descendants depicted as the younger outer ring, and a few fossil remains (the individual species inside the ring). The scattered life forms around us today (step 2) are just a fraction of all the life forms that have existed on Earth. We humans have tended to line up those life forms with what we see as the most advanced form, namely, ourselves, as the end of the line, with what we see as more "primitive" life forms following behind (step 1). Although the evolutionary line we picture is nothing but an artificial construct. Maybe it is all Aristotle's fault for getting us to think that way in the first place (see figure 5). Better to see the entire picture of evolution expanding in all directions instead of just focusing on the small part that led to us.

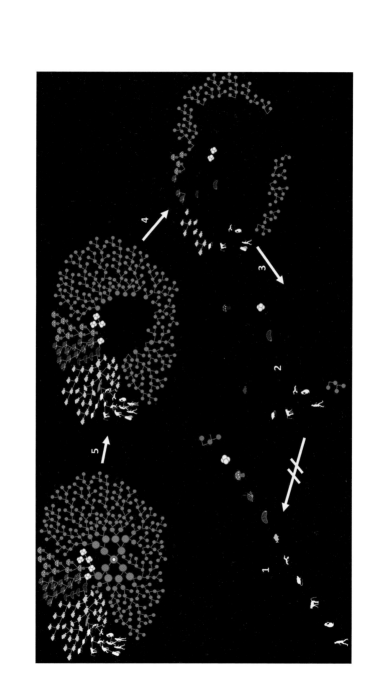

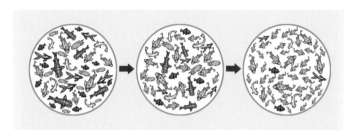

Plate 5 Natural Selection—Step 1: Make babies; Step 2: Babies struggle to survive because resources are limited, some die off; Step 3: Babies that prove to be best fit by surviving into adulthood make more babies like them. In this case the blue fish with spots are best fit and grow up to make more blue-spotted fish. Note that all the variation is not lost, a change in the environment may favor red fish with stripes or yellow catfish next season.

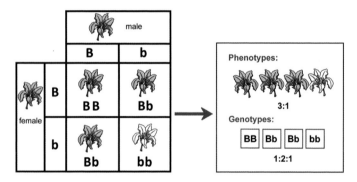

Plate 6 Crossing purple-flowered pea plants yielded three-quarters purple-flowered plants and one-quarter white. Why? Mendel deduced there were pairs of what we now know as "alleles" (different versions of genes that control traits like flower color) that separated during reproduction (the dominant B and the recessive b). Whenever the dominant B allele was present, a purple flower appeared, and whenever a recessive b was present without the dominant B in the combined seed, a white flower appeared. What Mendel's experiment showed us is how to separate the genotype from the phenotype to explain the maintenance of variation.

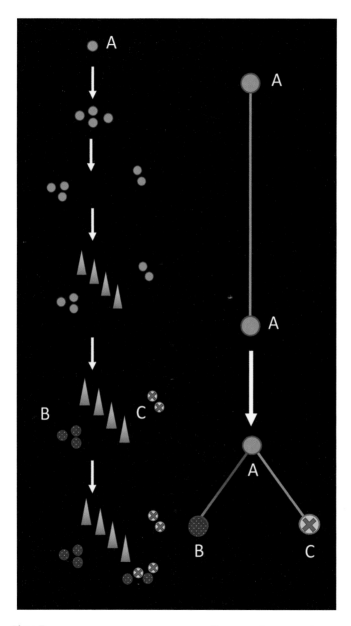

Plate 7

Plate 7   A hypothetical example of speciation. Individuals are shown on the left as circles, and how we might see changes in those individuals as part of evolutionary lineages are shown on the right. Species A, like all species, is made up of different individuals, and, like many species, is also made up of populations in different areas (left side). If separated by a barrier, such as a mountain range (blue pointy triangles) for a sufficient period of time, the members of a particular population may diverge into distinct species (B and C). The difficulty in determining when this might happen lies in knowing how much change is needed for speciation, which takes place somewhere along the lineages branching from A → B, and from A → C (bottom right). The changes that accumulate during the separation of this population from the others don't need to be adaptive—they can be neutral changes to the DNA. Eventually, enough changes may accumulate to reduce the likelihood of gene exchange even when members of the now-different species can mingle again (bottom left). These accumulated changes could also include things like new phenotypic differences (e.g., like color patterns or mating displays). In this example there are also new preferences in mates, with the blue X-shaped color pattern favored among members of species C, so that they don't attempt to breed with members of species B, which are purple and have spots instead. Or perhaps enough genetic differences arose between members of species B and C that even if they attempted to interbreed, they could not produce viable offspring.

SPECIATION: THE FORMATION OF NEW SPECIES

So, if symbionts don't force speciation in their hosts, as Lynn Margulis's "symbiogenesis" suggests, is there another formula for how a new species is formed? Yes: isolation + time. When part of a population becomes separated from the rest, over time, the unique gene pool in each of those separated groups will diverge both through nonadaptive changes (neutral mutations, genetic drift) and adaptive forces (natural selection), depending on the different environments those now-separated populations find themselves in. To begin with, since a smaller population will likely have less genetic diversity than a larger population, it will represent a smaller gene pool that can "fix" certain gene variants in, or lose them from, its members faster than a larger population can. This process is how the versions of the different traits that Mendel discovered, for example, become either one type or another over time. With fewer individual pea plants to cross with one another, you may end up, as Mendel did, with all purple-flower pea plants or all white-flower ones because of the genetic makeup of the smaller population. In fact, Mendel used artificial selection on a few individual pea plants to breed pure strains before he started crossing them for his experiments. In a real-life environment, there might be not only different sizes of populations, but also different types of predators or microclimates in the different areas where independent populations are found. These environmental differences may lead to different adaptations within the separated groups that may ultimately result in speciation.

There are usually multiple populations of any given species scattered around, be it ruby-throated hummingbird, death cap mushroom, wolf lichen, or any other species. Some of those scattered populations still interbreed with one another through regular migrations of their members together or the movement of a few individuals between one population and another. If, however, one of those populations becomes isolated from the others—let's say by a river that changes course and divides one land area into two—interbreeding between members of the more isolated population and members of the other populations will be limited because of the new geographic barrier. Given enough time, the accumulated changes within the different populations will build up to the point that even if the river returns to its original course, the members of the different populations might have changed enough that, no longer recognizing each other as belonging to the same species, they no longer mate. In that case, they would mate only within their own group despite the removal of the barrier (figure 10). Thus a new species is born.

Speciation due to isolation caused by the formation of a new geographic barrier is called "allopatric speciation." It is thought to be the most common way by which new species arise, but "isolation + time" may happen in different ways even without a geographic barrier. For instance, "sympatric

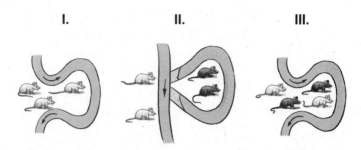

Figure 10 A meandering river (I) has a population of mice adjacent to and partly surrounded by a bend in the river; the river eventually makes an ox-box lake (II); the population of mice on the island formed by the ox-bow lake is now cut off—isolated—from the rest of the population of mice. It is slightly more humid and cooler on the island, and there are fewer hawk predators but also less food and fewer nesting spots. It is also muddier, leading to selection for darker fur to camouflage better. Eventually, the river returns to its original course (III) but not before the island mice have accumulated enough behavioral, physical, and genetic changes to be considered a distinct species.

speciation" happens when there are no physical barriers and little time, and it is more common in plants than in animals. Even adjacent individuals of similar external appearance in the same population of plants may develop duplicate sets of chromosomes ("polyploidy") instead of the typical number. Plants with different numbers of chromosomes usually can't interbreed since their chromosomes can't all align when their gametes come together. But they can interbreed when the number of their chromosome pairs is the same—three pairs with three pairs, four pairs with four pairs, and so on. When individual plants with the same number of chromosomes but a different number from most other individuals in the area pair up—boom, a new species may arise.[1] Sympatric speciation may have even happened in extinct human species of the genus *Homo* because a change in their immunochemistry may have caused a chemical fertility barrier—giving new meaning to two mates having "bad chemistry."[2]

But what is the definition of "species," you might ask. Well, the exact definition is complicated because living things on Earth have varied so much in the past and continue to vary so much even now that scientists can't all agree on a definition of "species" that works *every time* for *every species*, from paramaecia to dinosaurs. We likely won't ever have a single all-purpose definition of species—a situation most of us scientists find very unsatisfactory (unless you are a quantum physicist: check out their particle-wave duality, the uncertainty principle and multiverses). Biologists, in particular, want clear boundaries between species, like the boundaries between most countries. But, instead, the distinctions between species are often blurred like the shifting boundaries of deserts, for example.

As a general rule, if organisms of one population can no longer interbreed with similar organisms of another population and produce viable offspring, biologists say that they no longer belong to the same species. But there are many exceptions to this "rule." Seldom do we check if different species of organisms can actually interbreed. One way we do check, however, is by taking note of any hybrids, although the ability to hybridize doesn't automatically mean the hybridizing organisms actually belong to the same rather than to different species. Along with many other biologists, I was flabbergasted to learn in 2020 of the "sturddlefish'—an artificially crossbred hybrid of a sturgeon and a paddlefish, two groups that diverged from each other nearly 200 million years ago (longer than kangaroos and humans have been on separate diverging lineages).[3] It just goes to show that hybrids, even viable ones, don't always help scientists distinguish between members of different species—or even between members of different taxonomic families, as it turns out.

In practice, and to be practical, taxonomists use differences in heritable traits (e.g., blue scales versus red scales), often combined with genetic evidence that lineages are distinct, to name and describe species. However, paleontologists describe thousands of species a year but must rely on different physical features in limited and often incomplete fossil samples, and, in studying microbes, microbiologist rely on a suite of differences they expect to occur between species that might not work for vertebrates.[4] Despite their different approaches, however, biologists generally agree that species are real-life categories, not artificial ones invented for our convenience, and that species are the products of evolution, even if they sometimes disagree about when there are enough accumulated differences to make for "distinct species" rather than "independent populations" (figure 11).

The Galapagos Islands are considered the "birthplace of evolution" because Darwin recognized (with a lot of help from taxonomists and years after he returned to England) that the groups of animals (especially the tortoises and birds) on those separate islands were all slightly different from one another, even though each group likely had a single origin from mainland South America.[5] The animals on the different islands differed because each island was environmentally

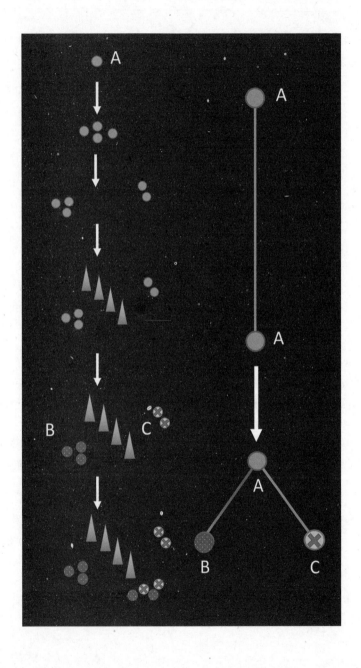

distinct and had different habitats and occupants that gave rise to different adaptations. Others would recognize (much later) that some populations differed even within particular islands as seasonal drying or cooling made some types of food more abundant than others, so that the struggle for survival shifted with time, as did which organisms were "fittest" (reproductively most successful). The most famous examples of these fluctuating populations on the Galapagos are now

Figure 11 A hypothetical example of speciation. Individuals are shown on the left as circles, and how we might see changes in those individuals as part of evolutionary lineages are shown on the right. Species A, like all species, is made up of different individuals, and, like many species, is also made up of populations in different areas (left side). If separated by a barrier, such as a mountain range (pointy triangles) for a sufficient period of time, the members of a particular population may diverge into distinct species (B and C). The difficulty in determining when this might happen lies in knowing how much change is needed for speciation, which takes place somewhere along the lineages branching from A → B, and from A → C (bottom right). The changes that accumulate during the separation of this population from the others don't need to be adaptive—they can be neutral changes to the DNA. Eventually, enough changes may accumulate to reduce the likelihood of gene exchange even when members of the now-different species can mingle again (bottom left). These accumulated changes could also include things like new phenotypic differences (e.g., like color patterns or mating displays). In this example there are also new preferences in mates, with the X-shaped color pattern favored among members of species C, so that they don't attempt to breed with members of species B, which have spots instead. Or perhaps enough genetic differences arose between members of species B and C that even if they attempted to interbreed, they could not produce viable offspring. (See plate 7.)

called "Darwin's finches."[6]* These examples of small evolutionary changes (ones that result in the small differences between species or populations that occur in shorter periods of time) represent what biologists call "microevolution." These small differences accumulate, leading to bigger evolutionary differences that we call "macroevolution," such as the evolution of flowering plants and associated insects.

------

* Even though Darwin made little mention of the Galapagos finches in his own writing (see endnote 6 to this chapter), the finches would subsequently become a model example of adaptive evolution, thanks to other biologists back in England who helped Darwin recognize that the little birds he so carelessly labeled (not even keeping track of which island each bird was from) were actually different species of finches. Even so, Darwin did recognize that the finches would have had a single ancestor from the mainland. Read about the fine work of Rosemary and Peter Grant on these finches in *The Beak of the Finch: A Story of Evolution in Our Time* by Jonathan Weiner. Thanks, also, to their habit of stealing food from tourist traps on their islands, some Galapagos finches may be evolving to prefer French fries.

ON FOSSILS AND THE BOOK OF LIFE

Darwin was criticized, fairly, as he noted himself in *On the Origin of Species*, for lacking much evidence from the fossil record to illustrate the gradual evolutionary change he expounded upon.[1] The fossils you would expect if life evolved in small steps, from one to another, were poorly known in Darwin's time. What was known then (and now) is that the vast majority of species that have ever lived went extinct without leaving any fossil remains.[2] In this chapter, I update and expand a metaphor Darwin and others used of an old tattered book to explain the incomplete nature of the fossil record.

If the history of life on Earth were recorded in a book— say, a great Book of Life (figure 12)—where each page was a species, most of the pages would be torn out, some would be unreadable, and some would be shuffled around. We have a good idea of such a book's larger sections—its "chapters"— representing the major groupings of animals at the level of "phyla." But most of what we know from the past is what we can read only from the living descendants that walk, swim, crawl, or fly among us today—as represented on the last few pages of each chapter's subsection at the level

Kingdom: Animalia
Phylum: Chordata
Class: Mammalia
Order: Primates
Family: Hominidae
Genus: *Homo*
Species: *sapiens*

Figure 12 The old tattered Book of Life. Many pages are missing including the introduction that explains the origin of life. Some pages have been shuffled around, and the book remains largely incomplete. Living species are well represented and they make up the pages at the end of each subsection that reflect the major classes of living things on Earth. The major, somewhat arbitrary taxonomic groupings, from largest to smallest, are: kingdom, phylum, class, order, family, genus, and species. We humans, *Homo sapiens*, are the last remaining living species in the genus *Homo*, which includes nearly a dozen extinct species, such as *Homo neanderthalensis*, the Neanderthals. The other members of our family Hominidae include chimpanzees and orangutans, and our order Primates includes monkeys and lemurs. At a higher taxonomic level, we are members of the class Mammalia along with other mammals such as dolphins, platypuses, and wallabies; as members of the phylum Chordata, we can identify with hagfish and sea squirts and all vertebrates that at some stage have a nerve cord, a notochord, pharyngeal gill slits, and a postanal tail; and finally, we belong in the kingdom Animalia with all the other animals. Some people also use the even broader category "Domain" for the three major division of living things: Archaea, Bacteria, and Eukaryota (see figure 4); we animals belong to the latter group along with all multicellular organisms.

of "classes" or "orders"—pages that remain in relatively good shape and that allow us to build the narrative of the entire book.

Some chapters of the Book of Life are in better shape than others, clearly influencing our view of life on Earth. Very few microorganisms leave any evidence of their existence, although it should be noted that the oldest evidence of life on Earth is from microorganisms that lived more than 3.5 billion years ago (stromatolites left by cyanobacteria and microbial remnants from hydrothermal vents).[3] So we have a rough idea how long ago the book's narrative of life on Earth started: since we can date the birth of our planet with geological evidence to 4.5 billion years ago, we know life took less than a billion years to appear. Unfortunately, the first chapter of the Book of Life is lost; the first living things left no "Introduction" for us to read.

Most of the earliest pages of the chapters of the Book of Life are biased toward the kinds of organisms that lived in environments that permitted fossilization and ones that had hard parts (like bone or shell) to preserve. When they die, the bodies of most organisms that live in humid, hot environments on land are consumed either by decomposition or by scavengers before they can become fossils. Marine fossils are far more abundant than land fossils; to make a good fossil, an organism's body usually needs to be buried quickly in an oxygen-scarce ("anoxic") environment. The sea provides far greater opportunities for quick suffocating death than land does.[4] Vast gaps and biases remain in the fossil record, however, despite the millions of fossils we have found since Darwin's day. In fact, more fossil than living species are being described every year, with, on average, between five and ten fossil species described every day.[5]

That wasn't the case when Darwin was alive; people back then were still arguing over what these strange stone forms represented. What fossils did for many at the time was prove that at least some species had gone extinct. It seems obvious now, but, for a long time, people thought all the life on Earth was all the life that had ever existed and that extinction was impossible. It became harder to believe this, however, when fossil hunters like Mary Anning (1799–1847) were discovering not just fossils of small, shelled creatures (Anning is often linked to the "She sells seashells by the seashore" tongue twister), but fossils of huge swimming reptiles and flying beasts like the pterosaurs and ichthyosaurs Anning was the first to discover.[6] Anning's discoveries convinced a lot of folks that these fossilized creatures had lived long ago but no longer existed: Extinction was real.

The fact that some of "God's creatures" went extinct deeply challenged people's religious views.[7] But this fact didn't exactly lead people to accept evolution. Instead, they came up with an alternative—there were periods of "special creation" where the Creator periodically destroyed and repopulated the Earth—a position sometimes called "catastrophism." And, as a matter of fact, there have been five major mass extinction events that were catastrophes for much of life on Earth. But there is no evidence of a wholesale repopulating of the Earth with independent, new origins of life, as "special creation" would require; rather, all the evidence points to the planet being repopulated by the survivors of extinction (just as mammals replaced the nonavian dinosaurs as the dominant group on land after the Cretaceous mass extinction some 66 million years ago).* "Special creation" was the leading view of many of Darwin's scientific nemeses, including America's most prominent scientist at the time, Louis Agassiz.[8] Agassiz had no small role in hardening the resistance of many in the United States from accepting evolution, and his legacy influenced the current creationist and eugenics movement.[9†]

Darwin himself discovered some fossil species while in South America, and these convinced him that not only did species go extinct, but they also evolved from one another. He found giant armadillo and sloth fossils in South America that he could identify as being relatives of the living animals he was seeing as much smaller forms in the same areas.[10] He deduced that the environment they lived in must have been very different from what he was witnessing around him.[11] He saw an evolving Earth and organisms that evolved with it.

---

* Although there are many excellent publications on this topic, I think Lauren Sallan's 2017 TED Talk is the best explanation of how life today is made up of the winners of past mass extinctions: https://www.ted.com/talks/lauren_sallan_how_to_win_at_evolution_and_survive_a_mass_extinction.

† Agassiz was an inveterate racist, and his disciples include David Starr Jordan who played a pivotal role in creating a eugenics movement in the United States and Europe. Jordan wanted to eliminate the "unfit"—the poor, handicapped, alcoholics, and many others. You can read more about Jordan in Lulu Miller's *Why Fish Don't Exist*. Jordan's academic tree includes almost every ichthyologist in the United States including me, but we inherited only his love of taxonomy and fishes, and the racists traits were lost (I hope). Members of the modern creationist movement are, for the most part, motivated by differences in religious perspective rather than scientific philosophy, but that may have been the case even for a nineteenth-century scientist like Agassiz as well.

Darwin was heavily influenced by the idea of "uniformitarianism"—the idea that the same geological processes that happen today happened in the past, and these slow and gradual changes build up over time to result in the major changes and structures (mountain ranges, canyons) that we see today.[12] Darwin himself witnessed an earthquake in South America that shifted the land in a way that made it easy for him to see that, over long periods of time, small and even sudden geological changes could add up to major landscape changes.[13] Darwin's theory of gradual change by natural selection is essentially the co-opting of a geological idea (uniformitarianism) for biology. The small microevolutionary changes in populations that he observed (such as in tortoises and finches with slightly different adaptations for feeding) would lead, over millions of years, to the macroevolutionary differences he saw represented in fossil forms and across the great profusion and variety of living things he called "the tangled bank."[14]

The reason Darwin was apologetic about the lack of fossils in *On the Origin of Species* was the expectation that there should be at least some true "transitional" or "intermediate fossils"—fossils of once-living creatures between a bird and a featherless reptile, between a whale and a terrestrial mammal, or between a fish and a land animal. But by 1859, few such fossil had been found (fossils couldn't be excavated as they can now, which makes Mary Anning's work all the more remarkable). Darwin needed these fossil intermediates to prove to the "special creation" believers that life evolved in a continuum of change and was not the result of separate periods of creation. The first of these remarkable intermediate fossils was first described in 1861, just two years after publication of the first edition of *On the Origin of Species*, although the significance of the discovery would not be fully appreciated for some time. This fossil, now one of the most famous in the world, was that of *Archaeopteryx*, the so-called "first bird," which had reptilian features such as teeth and claws, but also feathers. This discovery was a shocking find at the time because feathers were thought to belong only to birds, and this creature looked much more like a dinosaur than a bird (because it is both).[15] Unfortunately, another Darwin rival, Richard Owen, obtained the fossil *Archaeopteryx* and, for years, kept it from being recognized for what it was (a transitional fossil) out of what was likely professional jealousy. Darwin had a complicated history with Owen, who had coined the term "dinosaur" and had written early on about "transmutation of species," but who would later write a scathing "anonymous" review of *On the Origin of Species*.[16] Darwin, for his part, would retaliate by naming Owen in his "Historical Sketch," which made it into later editions of *On the Origin of Species,* starting with the third.[17] The sketch listed Owen among those who had influenced the formulation of Darwin's evolutionary theory, a theory Owen would have hated to be associated with.[18] That was quite the trick play from our old boy Charles.

Today we know of lots of intermediate fossils, including several other individuals of *Archaeopteryx* and many other feathered dinosaurs that helped us flesh out the dinosaur section of the Book of Life.[19] The book's section on birds was once thought to be independent of the sections on dinosaurs and other reptiles but we now know it is actually part of the subchapter on theropod dinosaurs—the birds we see flying about today are just living descendants of those extinct dinosaurs. Simply put, birds *are* dinosaurs.

We also know of intermediates between whales and terrestrial mammals, as we can follow the fossil record over time as a hippo-like mammal transitioned to become fully aquatic: we can see the nostrils successively move to the back of the head and become the blowhole and how the limbs became flippers and the tail a fluke.[20] We have an intermediate species Tiktaalik (*Tiktaalik roseae*)—the "fishapod"—that shows the transition between a fish and tetrapods (all vertebrates on land). Tiktaalik and other early lobe-finned fishes evolved thicker fin bones for fighting gravity while feeding in shallow water and had internal nostrils ("choanae") to allow them to breathe atmospheric air (and which is why you can breathe through your nose). We have a good series of intermediate fossil forms that were even more terrestrial than Tiktaalik based on their increasingly less-fishy anatomy,[21] and one species, *Qikiqtania wakei*, that said "nope" to staying on land and adapted back to a fully aquatic life.[22] We have intermediate fossils that help explain everything from how the turtle got its shell[23] to how we humans got our crappy bodies (more on that later).[24] Although we will never have a complete Book of Life with all the pages restored, we have enough of the pages to connect major points of the story of life on Earth. If we had all the pages, we could see the complete view of life with all the branches of every creature that has ever lived shown instead of just the incomplete Tree of Life we have representing what we do know.

The pages in the Book of Life aren't numbered, of course, but DNA analyses of living species help us connect who is related to whom in a way that gives us at least the outlines of the book's chapters and much of its index. We also know the Book of Life isn't read in chronological order from beginning to end. Each chapter is still evolving and changing, and all its parts may be on independent trajectories. Just because one fish came onto land doesn't end the aquatic fish chapter— they are still evolving and diversifying in our oceans, rivers, and lakes. Likewise, just because some mammal lineage gave rise to humans doesn't mean the sections on rodents, bears, or even other primates (in the mammal section of the Chordata chapter) don't continue to add new pages (species). The ends of each section in the book are the recent species; the only finished sections are those where all the species went extinct and left no living descendants. Sadly, there are many of those, and some of those endings are still being written today, because of our actions.

III

QUESTIONS AND
MISCONCEPTIONS

WHO YOU CALLING "PRIMITIVE"?

Do a little navel gazing. You have a belly button because your mother was a placental mammal. You developed in her uterus with the placenta providing oxygen and nutrients to you through the umbilical cord, which was cut when you were born and left you with a useless skin-plug remnant of that outlet. Mom was left with her organs squished, her body bruised and a hungry little monster that always wants her milk and that won't let her sleep.

Consider the kangaroo, a large biped—just like us except aided by a strong, steadying tail. The mother kangaroo gives birth to a tiny fetus that comes out half-formed from one of her three (yes, three) vaginas.[1] The fetus crawls up on its own into her snug marsupial pouch with nipples in it, where it will stay and feed off its mother's milk until it can strike out on its own. To me, that sounds a lot more comfortable than the way we humans come into this world. Our large brains require even larger heads, and human babies have to be squeezed out in much more precarious circumstances than our kangaroo cousins do. We need to be born long before we're fully developed and able to fend for ourselves because if we waited longer, our heads would be too big to fit through the birth canal which makes us very needy over our first year of life and pretty needy for years after that. Can you think of any other animal newborn that can't even sit up and that can barely keep their food down over the first six months of their lives? Every time I hold a newborn baby I think, "This kid could sure use more time in the oven." At least the kangaroo fetus has the good sense to go back inside the mother for another round of gestation. (See Holly Dunsworth's work for an alternative take about human births, although I don't think anyone would argue against the idea that a marsupial birth, like in the kangaroo example, is more comfortable and safer than a human one.)[2]

We tend to call marsupials like kangaroos "primitive"—but they shouldn't be called that, nor, for that matter, should any living things around today—they are all survivors and parts of the budding leaves of the Tree of Life, where marsupials are the "sister group" to us placental mammals. Our lineages are of the same age and diverged from the same ancestor at the same time, in other words, we are "closest relatives."* That means all us placental mammals share a more recent ancestor with the marsupial lineage than we do

---

* I know what you are thinking—why not just say "closest relatives" then, *without* all the technical language? Well, usually I try to avoid technical language, but, in this case, I want to make it clear how saying "sister groups" better explains the uncertainty built into looking for ancestors on the Tree of Life and to show how biologists leave room for discovering new relationships in the present and the past. The real "closest relatives" are those undiscovered ancestors equidistant between those "sister groups" that are like twins from the same mother. Because we don't know the ancestral mother in most cases, we can call the descendant twins "closest relatives" but "sister groups" is technically more accurate.

with the third major group of mammals, the "monotremes," such as the platypuses and echidnas.[3] But before you call the egg-laying platypuses "primitive," think of all the amazing features and abilities they have. Their duck bill helps them hunt even in murky waters by detecting the electric field around their prey, their conveniently webbed feet help them swim faster and longer. Male platypus even have a venomous spur on each hindleg to help them defend themselves. And the females are able to nurse even though they have no nipples—by releasing their milk through pores on their skin where their babies can lap it up.[4] Oh, and in 2020, we learned that platypuses are one of just three known mammal species that are biofluorescent—so they glow under ultraviolet light.[5] If a platypus were looking at you, a soft mushy human burning in the sun, they would think they were the advanced ones. It is slightly better to just call some feature of an animal "primitive" instead of the entire animal, like the egg laying in monotremes because it also occurred in our evolutionary ancestors. (In systematic biology we call that kind of trait "pleisiomorphic" rather than primitive to emphasize that it is just a trait in the ancestral condition. Other people use "vestigial" in almost the same way.)

We don't call humans "primitive" just because we have so-called vestigial organs like the appendix or because we have wisdom teeth (you know those late developing teeth that don't actually fit in our mouths).[6*] In short, we shouldn't call any life form "primitive," or at least not wholly so. And, yes,

---

* Although the appendix was long thought to have no useful function, recent findings suggest it may act as an important repository for beneficial bacteria in the gut.

even some of those single-celled microorganisms like bacteria that you are thinking about have advanced abilities that we lack, like being able to digest and break down everything from wood to crude oil, or their ability to live through long periods of suspended animation—and even fight off viruses with their CRISPR-based immune system. (I'll explain that last one more in part IV; suffice it to say, they don't need to stay home because of a measly coronavirus pandemic.)

How about the humble fish? They were the first animals to develop backbones, jaws and the basic body plan of all vertebrates, because they *were* the first vertebrates. It was in the water that our fishy body plan with a big centralized brain, nostrils and lungs first evolved.[7] If, on the other hand, we had evolved from octopuses, we would have soft boneless limbs and multiple brains and hearts dispersed throughout our bodies. But we don't have these features because we evolved from a fish. And because the most recent common ancestor of all vertebrates was a fish, all vertebrates are technically fish. One branch of the fish lineage came onto land 400 million years ago: these are the lobe-finned fishes ("sarcopterygians"), which gave rise to Tiktaalik and other four-limbed vertebrates ("tetrapods") and to the roughly 35,000 living—and many more extinct—species of mammals, reptiles, birds, and amphibians. Other lobe-finned fishes were part of lineages that stayed in the water and that include living forms like the coelacanths and lungfishes.[8] Lungfishes do indeed have a lung and can drown if you hold them underwater. Coelacanths swim in a dog paddle and have thick limb bones like our radius and ulna that all land vertebrates have instead of the thin flexible rayed fins you think of in most fishes.[9] Coelacanths and lungfishes are more closely related

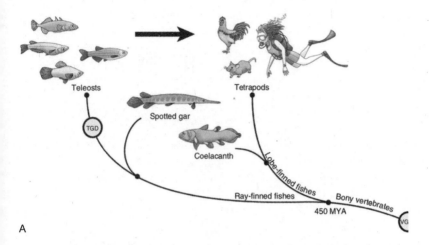

A

Figure 13 Images depicting the vertebrate (fish) section of the Tree of Life including four-limbed vertebrates ("tetrapods"). Note how we are more closely related to coelacanths in the lobe-finned fish lineage of both trees (see second tree on next page). Reprinted from Ingo Braasch, Andrew R. Gehrke, Jeramiah J. Smith, Kazuhiko Kawasaki, Tereza Manousaki, Jeremy Pasquier, et al., "The Spotted Gar Genome Illuminates Vertebrate Evolution and Facilitates Human-Teleost Comparisons," *Nature Genetics* 48, no. 4 (2016): 427–437.

to us (as we are also in the lineage of lobe-finned fishes) than to any other kinds of fishes. Therefore, a coelacanth is also more closely related to a human, like you or me, than it is to any ray-finned fish like a trout or a goldfish.

Going further back, you and I are also more closely related to a goldfish or a trout than to sharks, because we have a bony skeleton (like goldfish, trout and all bony fishes) and not a cartilaginous one (like sharks).[10] You might consider a shark a fish, and you would be right; but you might also think all fishes swimming around the oceans and rivers are more closely related to one another, but, on that point, you

B

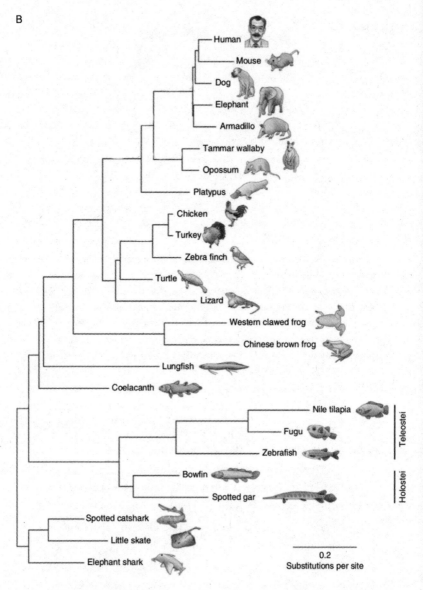

Figure 13 (continued)

would be wrong. Cartilaginous fishes like sharks have a more recent common ancestor with one another, and bony fishes are all more closely related to one another than to any other group. Another way to think of this is that the class of cartilaginous fishes has a different section in the Chordata chapter of the Book of Life than the class of bony fishes do (see figure 13).[11] The cartilaginous fish section, has fewer species pages than the bony fish section, and the bony fish section includes not only tuna and tarpon but all land vertebrates, because again all four-limbed vertebrates descended from a common fish ancestor, including humans. So yep—you are a fish my friend—we humans are all just fish out of water—and that's why our bodies are a flop.

OUR CRAPPY-CRAPPIE BODIES

When it comes to how they're built, our human bodies are something of a disaster. I don't care if you look like Beyoncé or Bruce Lee. Our bodies are hunks of water-logged flesh wrapped around collagen and calcium sticks, held together by strings of blood and muscle with a twisty tube of bacteria running through the center, all covered in an oily skin bag. We are frail naked apes with oversized lollypop heads so ill equipped for life that we get tired after standing still for ten minutes. Why? Well, again, because we are literally fish *out* of water.

We are taught to think we humans are perfect—the "*pinnacle* of evolution," no less. Hogwash. The only human feature worth going on about is our big brain, and we use that brain just enough to think we are better than every other living thing, and to build stuff that makes us feel important—but that may also destroy all life on Earth, ourselves included. So it's time for some humility.

Take our knees, for instance. Why aren't they ball-and-socket joints like our hips and shoulders are—or at least something simpler than an oddly shaped bony wedge with a network of rubber-band like tendons and ligaments attaching it to the bones of our legs that all end up rubbing together painfully in old age?[1]* Why do we walk upright? And why do we get so tired just standing? You are probably sitting right now or wondering when you'll get to sit or lie down next. That's because we are constantly fighting gravity. You know who doesn't have that gravity problem? A fish *in* water.

Consider the teleost: these advanced bony fishes (teleost means "perfect bone") are the beautiful product of 500 million years of fishy evolution.[2] The 20,000 teleost species are the real pinnacle of bony fish evolution (these ray-finned fishes have nothing to do with the lineage of lobe-finned fishes that broke away on their own 400 million years ago and that came onto land and, eventually, made you and me).[3] Most teleosts have a balloon-like swim bladder that lets them float around leisurely, making them neutrally buoyant in the water. I am thinking of a seahorse at rest lounging around with its tail holding on to a leafy plant in the sea, like a jaunty tethered astronaut floating in space.

Teleosts have neat little spool-shaped vertebrae that are loosely connected to each other in a nearly straight line; this "backbone" is strong enough to let them propel themselves through the water with a flick of their tails without having to worry about hurting their backs. Their hearts are up close to their heads, so they can get oxygen quickly through their gills and then pump that blood around their bodies with the low blood pressure of a tranquil bon vivant. Even if they

---

* I really like Jay Olsansky, Bruce Carnes, and Robert Butler's description of how to build a better human in their 2001 article "If Humans Were Built to Last." It ain't pretty though. Not as pretty as a teleost, anyway.

get injured, they can recover quickly, and some teleost fishes can regenerate injured heart, brain, and spinal cord tissue.[4]

Instead of perfecting the early fish system for the aquatic world, as teleosts have, our human ancestors had to take the early lobe-finned fish system and make it work for life on land; and we've suffered the consequences ever since. We have higher blood pressure to pump against the force of gravity, putting a lot of strain on our weak hearts, which have just a pair of coronary arteries feeding them blood. If a single artery gets clogged, we may die (as many people do every day when they have a "coronary"). Standing upright makes our blood pressure worse, since we have to fight gravity more when we stand. That's why it's recommended we lie down when we need to lower our heart rate (like after a venomous snake bite or a coronary). Fish hearts are a lot more foolproof, being fed by big pools of blood that don't get clogged so easily just because they ate too many fatty jellyfish sandwiches.

Our odd land-based circulatory system made it not just harder to oxygenate our blood but also led to the evolution of one of the strangest parts of our anatomy: the laryngeal nerve. That nerve supports the larynx or "voice box," which evolved from the arches in the throat or pharynx (pharyngeal arches), derived from the gill arches of our fishy ancestor.[5] The vagus nerve connects the brain to the heart and helps determine the heart rate and speed of respiration in fishes (via the sixth pharyngeal arch).[6] In humans, over the course of our development, the larynx descends down the neck and is suspended by a hyoid cartilage (which fishes also have for supporting their gills), and this voyage down the neck pushes the throat's (laryngeal) branch of the vagus nerve down toward the heart and back up the throat in a roundabout loop (instead of a direct path between brain and throat). The circuitous route taken by this nerve would be true of any four-limbed vertebrate and may even have approached 100 feet in long-necked sauropod dinosaurs like Apatosaurus or Dreadnoughtus.[7] That's a lotta nerve.

It isn't all bad. We can get more oxygen from the air as humans than our ancestors could from the water as fish. But even though water has a lot less oxygen than air does, gas exchange is more difficult with lungs than it is with gills. Worse yet, we humans use the same tubing for breathing as for feeding, and we have just a little piece of flappy tissue (the epiglottis) to keep food from going "down the wrong pipe," which, of course, it often does.[8]* When all is said and done, I'd rather have the fishy body of a crappie than the crappy body of a human.

Fishes also keep their baby-making equipment snuggly inside their bodies. Not so for us humans. From boys to men, we have our meat and potatoes hanging outside our bodies because we produce too much heat inside them (which would kill the developing sperm), But, of course, our testicles have to start off inside our bodies because that's where our "cold-blooded" ancestors had them. As they drop in our bodies, the testes drape over the ureter in a half-knot that

---

* Or, as Bill Bryson puts it in *The Body: A Guide for Occupants* (2019): "We were built to choke." We have longer necks than other primates probably as a function of our bipedalism, which may have also endowed us with the ability to speak, which also gave us the ability to choke—it's all a circular mess. Notably it seems it was actually a loss of morphological complexity that freed us to evolve speech (see references in note 8).

causes prostate issues later in life and potentially hernias early in life.[9] TMI Fun Fact: I had an undescended testicle that required an operation as a child—maybe that's when I started hating the human body.

We can see in our bodies the evolutionary connections we have with more recent ancestors, too. We still have the remnants of a tail (our "coccyx," which is just an extension of our spine), and when we get goosebumps, it's to raise the hairs of the fur we lost when we became "naked" apes. Yes, we had hairy ancestors with tails, but, no, not from monkeys. We did not evolve from monkeys. We share a common ancestor with monkeys that led to both tailed monkeys and to the tailless great apes (of which we are one).

"Where are we from?" To truly know—and understand—the answer to that age old question, we need a map—a map called the "Tree of Life" and one we should learn how to read to understand our origin as a species.

THE TREE OF LIFE

In studying the evolutionary relationships between species, my main tool is the Tree of Life, which displays the relationships between all the species that have ever lived on Earth. This representative tree is a way of visually describing how species are related in a series of ancestor-to-descendant relationships (go back to figure 4). I focus on the part of the tree related to animals, specifically the part related to vertebrates, and even more specifically, I study fishes. If you zoom in on the fish part of the tree and the roughly 35,000 living species (and all the extinct ones we know about), you'll see that the fishes represent only a small fraction of life on Earth.[1] There are more than 350,000 described species of beetles alone, more than any other animal group, and almost the same number as that of all described plants—but about ten times the number of described bacteria.[2] Part of the reason for that surprisingly lopsided difference in species richness is that there have been historically more entomologists describing beetle species than microbial taxonomists describing species of bacteria. There are almost certainly many more species of bacteria, than of beetles, but these microbes are also much harder to describe taxonomically.[3] Perhaps if bacteria were the size of beetles, things would be different. In recent years, we've been able to sample a wide range of environments using environmental DNA and at least "guesstimate" the number of new microbial species in any given environment,[4] but identifying their presence is quite different from identifying, naming, and describing those species for science.

The Tree of Life tells us about more than species diversity—it also tells us how the organisms in any group of species are related to one another and which groups of species are "real." A group of species is "real"—or a "natural group"—if it includes a common ancestor and all the descendants of that ancestor, for example, the common ancestor of horses and all horse species descended from that ancestor that have ever lived or gone extinct. The Tree of Life can also tell us when a group of species is "not real" or "not a natural group." And, as it turns out, some of our favorite groups are "not natural": lizards (when you don't include snakes) or raptors (owls and hawks and other birds of prey that kill with taloned feet), or even the great apes (when you don't include humans).

Sadly, even my favorite group, "fishes," aren't a "natural group." I know, I also felt lied to the first time I realized this, but the situation is complicated. Because fishes include both cartilaginous and bony species, fishes are only a natural group if you place the shared common ancestor of both groups at the root of all vertebrates. So, unless you can call all vertebrates "fishes," it isn't a natural group. Same thing happens if you remove the four-limbed vertebrates ("tetrapods") from the fish lineage. Removing members of a natural group makes it "*un*natural" because then it would not include all the descendants of a common ancestor and, therefore, wouldn't reflect the true evolutionary history of the entire lineage. It would be like not counting your grand-mother as part of your family because you've disowned her—even though, biologically, you and she still belong to the same family. So, unless you admit you are a fish, along with snakes and pandas, salmon and peacocks, you are say-ing you belong to an "unnatural" group—and you are break-ing up your "real" family (go back to figure 13).[5]*

Our understanding of the Tree of Life changes daily as new relationships are proposed, new species are described, and new data becomes available. Most of the tree remains stable, particularly the largest and deepest branches near the trunk since most of the changes are at the tips of the outer branches.

The study of the relationships between living things on the Tree of Life, which is called "phylogenetic systematics" (or just "systematics"), has changed dramatically during my career: most active systematists studying living spe-cies almost exclusively use DNA to determine relationships between them. As more data becomes available, especially genome-scale data, our understanding of the Tree of Life is updated, and the tree itself evolves, just like the living things and their relationships it is describing. Check out the "Open Tree of Life Project," which uses the tree and leaf metaphor for the continually updated Tree of Life.[6]

Even though the Tree of Life represents ancestor-descendant relationships, it does not have any ancestors named on it. All the branch tips on the tree are species, with

* My TED Talk on evolution covers the "you are a fish" subject as I like to tell it, "Four Billion Years of Evolution in Six Minutes"; it also inspired me to write this book.

the ancestors represented by the nameless "nodes" or bases of the branches. We evolutionary biologists never say, "This species is the ancestor of these species," unless we're talking about wolves being the "ancestors" of dogs in that we know dogs are descended from them through human-directed selection (breeding) that we've largely witnessed. We don't look for ancestors, because ancestors are generally "unknow-able"—it is unlikely that we stumbled upon the individual, population or even species that gave rise to a new lineage. And how we could even recognize an ancestor as such is not always clear. What we are really looking for are "sister lineages"—groups that are each other's "closest relatives," and that evolved from a "most recent common ancestor" together. These sister-group relationships can help us explain how evolution happened in more precise language. We can say, "These sister lineages split off when South America and Africa broke apart, or when this mountain range formed or when one of the lineages had a duplication event in its genome."

We biologists develop hypotheses of relationships based on modeling the evolution of heritable characters. We look at which features are shared and were modified by a group of organisms: the transition of scales to feathers, for instance, defines one group of theropod dinosaurs, the group that includes birds. Likewise, the transition of a single DNA base from A to C (adenine to cytosine) might be evidence that the bacteria in a particular group are each other's closest relatives (although we generally have much more evidence than that to determine whether they actually are).

Although the Tree of Life is a map of the Earth's evolution-ary history, it will always be incomplete, not only because we don't know all the living species there are to know, but also because we certainly don't know most of the 99 percent of species that have gone extinct in the nearly four-billion-year history of life on Earth.[7] The life span of most species is short, at least in evolutionary terms. The species we see alive and around us today are all about the same age as our own species, *Homo sapiens*; we living species are all about a few million years old or younger.[8] And we living species are all about equidistant from the center, the "Big Birth"—the birth of the first life form, or at least the most recent common ancestor of organisms that are alive today (go back to figure 6). Superficially at least, some living species look like some of the earliest living things on Earth, and, in fact, for about two billion years most of life looked, well, pretty much the same—a seemingly endless number of tiny individual single-celled creatures.[9]

But despite their unassuming appearance, there was and is an astonishing diversity of species within that unicellular life. Some living things that flourish in extreme environments today probably look like the Earth's earliest life—we call these organisms "extremophiles" and many are members of the domain Archaea.[10] They live in near-boiling water or hydrothermal vents or other inhospitable places that may have resembled the Earth around four billion years ago. Other unicellular life incorporated mitochondria and chloroplasts; which resemble other organelles in our cells but lived independent lives on their own before they were incorporated.[11] Unicellular organisms without nuclei are members of either the Archaea or the Bacteria; all multicellular organisms with nuclei and other organelles are called "eukaryotes" and are members of the Eukaryota—including us.[12] (And, yes, there are, of course, also unicellular eukaryotes[13] because evolution does not take to being boxed in.) But how did all of these living things come to be in the first place? Well, we don't know. But that's okay: admitting and confronting our ignorance is where we get our power as scientists. From doubt and ignorance, we get curiosity, and from curiosity we get conjecture, observation and experimentation. And from all that we get answers, or so we hope. But even the answers we do get are just hypotheses—we need to challenge them with new questions and to test them again and again. And that's part of why we don't know how life originated with any certainty—at least not yet.

We do know that life gives rise to life. You and I and all other creatures living today were always alive through the countless ancestors before us. And so you and I are part of a chain of unbroken living forms that goes back to the common ancestor of all living things. There were likely living things before that, but perhaps they had less successful descendants that died off before leaving more successful descendants of their own. But how did that first life come to be? And, for that matter, what *is* life? Defining "life" or "living" is actually pretty hard. Many still use something close to Aristotle's 2,000-year-old definition of something that is capable of "self-sustenance, growth, and reproduction."[14] But some organisms gain nourishment only from what bacteria digest for them, such as the mouth-less giant tube worm *Riftia pachyptila*;[15] many unicellular organisms don't grow; and some living things can't reproduce, such as the sterile worker bee. In the next chapter, I'll alter Aristotle's definition to include an evolutionary perspective.

THE ORIGIN OF LIFE

There is no knowledge without uncertainty. However, discouraging that statement might seem, there's much truth in it. But perhaps a more useful statement (and a good motto for scientists) might be one attributed to Mark Twain: "It ain't what you don't know that gets you into trouble. It's what you know for sure that just ain't so."

Where did we come from? Other life. But where did the first life form come from? We just don't know. We don't know because we don't have a fossil of the first living thing, nor do we have a fossil of the "intermediate" between the nonliving (inanimate matter) and that first living thing—the ultimate transition. Not that we haven't looked. Just as with *Archaeopteryx*, scientists thought they had discovered an important transitional form shortly after the publication of *On the Origin of Species*. "Darwin's Bulldog" Thomas Henry Huxley thought he had discovered a "protoplasm" that was potentially the transition—intermediate—between the nonliving and the living. He named this odd, apparently animate so-called protoplasm retrieved from the Atlantic seafloor *Bathybius haeckelii*. Unfortunately, what Huxley thought was an example of a very simple intermediate, proto–life form turned out to be no more than an artifact of staining and preserving seawater.[1] To his credit, he owned up to his mistake. The eminent zoologist Ernst Haeckel "Darwin's Chihuahua" took longer to acknowledge Huxley's error, and perpetuated the myth that the would-be very first intermediate, which was named after him, was genuine. Few serious claims of discovering the greatest transitional life form have been made since.[2]

We are left with no direct evidence of how inanimate matter transitioned into animated life. But that's just how it sometimes is with science. Not knowing things with certainty is especially true in the "historical" sciences, where scientists must rely on information from the present to infer what took place in the past because, of course, they can't go back in time to make direct observations.

Scientists also haven't managed to produce life from inanimate matter in a laboratory or to observe it evolve on another planet. But their lack of success in doing either doesn't mean life did not arise on its own here or evolve elsewhere. You may not have met your great-great grandparents or even have pictures of them, but you know you had them and that they were human beings. It's too easy to write off the origins of life on Earth as "unknowable," but that isn't the scientific way.[3]* We have to keep searching and thinking, asking questions and looking for answers.

There are many different ways you could answer the question "Where did life on Earth come from?" Your answer

---

* If you want to get into the real details of the logic and philosophy of science and evolution, read, Elliott Sober's *Evidence and Evolution: The Logic behind the Science*.

or hypothesis could be "God created life on Earth," and I wouldn't be able to prove you wrong because there is no scientific test for "God did it." And if you stole your neighbor's car and told the court, "God made me do it," no one could prove you wrong there either—but, you'd still go to jail. So, let's stick to testable hypotheses about observable things in nature. With that in mind, there are two alternative hypotheses we can work with to explain the origin of life: (1) the "panspermia hypothesis": RNA/DNA-based life on Earth is extraterrestrial in origin and got here on a meteorite or by some other means; or (2) the "abiogenesis hypothesis": Life evolved on Earth from inanimate matter. Let's look at what we can observe in nature to test and support or falsify these hypotheses.[4]* In the end, both hypotheses might be right or both might be wrong—and there may be another alternative hypothesis we haven't thought up yet.

If you look at all life today, it is DNA based; therefore, the "last universal common ancestor" (LUCA) of organisms living today was likely DNA based as well.[5] Did life always have to use DNA? No—earlier life forms may have used the simpler RNA genetic code, or perhaps an even simpler replicating code before RNA, but all "standard-issue' living organisms today use DNA, at least all living species of organisms on the Tree of Life, since there are viruses that use RNA for their genetic code, although they still need DNA

---

* I like Karl Popper's "principle of falsification": Take a given hypothesis and try to prove it false, and if you can't, that doesn't make it true, just not "untrue." Keep testing the hypothesis in different ways. If you still can't prove it false, the hypothesis may be said to "stand the test of time," it is "corroborated" for now, but you must continue to test it rigorously.

intermediaries. Perhaps that first living thing on Earth was something like an RNA-based virus, but it may have been even simpler, something that left no descendants we can study today. We tend to define life as DNA based because all the living things we know about use a DNA genetic code. But what if we found virus-like biological entities on Mars that somehow could replicate on their own without DNA or RNA? If we discover potentially new life forms elsewhere, beyond Earth, we might even discover evidence of them having been on our own planet and realize we just didn't know what we were looking for because of our narrow definition of life.* Or maybe life was an inevitable outcome of the physical forces of nature?[6] We are clearly still in the "hypothesis-making" stages of that question.

So what is "life"? For me, "life" should be defined as "a property possessed by any organism that can inherit or pass down heritable traits and that can potentially participate in evolution (as part of a population susceptible to the neutral and selective forces of nature)." All living things we know of evolved and carry heritable information via a nucleic acid (DNA or RNA)—maybe that is all we need to think about to

---

* What is life? That is the ancient question that Aristotle struggled with in De Anima. Life is whatever separates the inanimate from the animate. But maybe the distinction between the "living" and "nonliving" is indefinable or even irrelevant. Maybe our real legacy as clumps of matter is how we transform energy and spread information? While we're at it, what is death? When we die, not all of our cells die all at once, and many of the cells that make us us, are actually our bacterial symbionts. So, when do we actually die? When all the cells with our DNA die? But what about the DNA we pass down to our children, or the DNA that our family passed down to us and that we share with them?

find the single source of life or at least the last universal common ancestor, which doesn't have to be the first living thing
ever, just the most recent thing that all living forms on Earth
alive today are descended from.

Let's first look more closely at nucleic acids. What does
DNA do anyway that is so special? Well, DNA is the genetic
instruction code for making amino acids (with help from
RNA), which make up proteins, and proteins make up you,
as well as the things that make you do the things you do.
You can think of yourself as essentially a vehicle for DNA
replication, that is, we organisms are just tools for DNA to
make and spread more DNA. As it turns out, considering all
they do, amino acids (those intermediates between DNA and
proteins) are surprisingly easy to produce. In one of the most
famous experiments of all time, Stanley Miller and his colleagues tried to replicate the early Earth's atmosphere within
closed chambers of gas, and when they then added heat and
electricity—voilà, they produced many of the amino acids we
observe today.[7] So that experimental outcome would seem
to support the abiogenesis hypothesis, except that making
proteins from exact sequences of amino acids is rather more
complicated. On the other hand, intact amino acids have
also been found on meteorites[8]—which would seem to support the panspermia hypothesis, although the unimaginable
vastness of space would seem to make that hypothesis less
likely to be true. And even if true, it just shifts the origin of
life from Earth to someplace else, without saying anything
about how life actually originated.

That said, it is notable that most of the water on Earth is
probably extraterrestrial in origin and older than our solar
system.[9] Where there is water, there may be life, and there

is water on or around our neighboring planets. In fact, there is much more water flowing or frozen on three of Jupiter's moons than on Earth, and even sizzling hot Mercury has ice. But that water occurs in many places in our solar system does not mean that there is life in those places.[10] News reports in 2020 of there being biochemical signatures of life on Venus recalibrated this debate.[11] And, even though those reports appear to have been mistaken,[12] at least they made us wonder about the signs of life we might need to look out for in future searches.

Putting aside for now the two alternative hypotheses of how life came to be on Earth, we have evidence beyond DNA and RNA for all life having a single common ancestor. As it turns out, all living organisms on Earth use left-handed optical isomers of amino acids instead of right-handed ones (think of them as left or right mirror images).[13] And all living organisms also use only right-handed sugars.[14] Right-handed optical isomers of amino acids and left-handed sugars are also available for synthesis and could easily have been the basis of life, but they were not and are not used by living things. That means all living organisms on Earth use the same genetic code and machinery for protein synthesis and metabolism. Neither you nor I can metabolize right-handed amino acids and left-handed sugars. And neither can an amoeba, a blue jay, nor, for that matter, any other living organism on Earth. (Somebody please make food out of right-handed amino-acid based proteins and left-handed sugars already. Get me that good "antichiral" cereal so I can snack unrepentantly.)

The fact that amino acids are easily made in the lab or sometimes found on meteorites isn't enough to explain how we get from these "building blocks of life" to actually evolving living things. We need something to copy the code that makes those building blocks (like how RNA helps make these amino acids or like what the enzyme polymerase does to DNA). Notably, scientists replicate DNA in labs across the world by using a polymerase chain reaction (PCR) with a polymerase first discovered in an extremophile microbe from a hot spring in Yellowstone National Park.[15] Why does that matter? Because hot springs are probably some of the closest environments we have today to what our planet was like when life was evolving over three-and-a-half billion years ago, shortly after the formation of the Earth.[16] And how do we know life on Earth may have been around that long ago? Fossils, of course. We have the layered mound "stromatolites" ("accretion structures")[17] formed from layers of ancient blue-green algae ("cyanobacteria"); we can still watch this accretion process today with modern algae. So, although we don't know exactly when life evolved or how, we do know it must have happened *by at least* 3.5 billion years ago, the time we start finding traces of life in the fossil record.

THE HISTORY OF LIFE

After the start of life, which must have been quite exciting, it gets boring for a billion years—"the boring billion": for millions and millions of years, all we have is just unicellular life evolving ever so slowly,[1] and most people just can't wait to get to dinosaurs—there is a reason there isn't a movie called *Proterozoic Park*. But even today there is plenty of unicellular life evolving: microbes, which still dominate the Earth—and dominate our bodies, too.[2] In fact, we wouldn't last long without microbes (most of the cells in our bodies are actually microbes, and many of our genes had microbial origins). Indeed, there's a good case to be made that we're just buses for these tiny "simple" creatures.[3]* In those first billions of years, microbes were learning how to do essentially everything on their own, so when we humans finally came along, they were ready to use us as just another inviting home to invade. Microbes are fully in charge of all essential processes

---

* Check out Sarah Hird's amazing (and funny) 2019 BAHfest talk on why we humans are just a convenient transportation device for the many microbes that control us. This may be the greatest science talk of all time: https://www.youtube.com/watch?v=6jGDPH5h_Mg.

of life, whereas many multicellular plants and animals rely on microbial symbionts to carry out critical life functions such as digestion on their behalf. In short, microbes can live without us—but we can't live without them.[4]

In those first few "boring" billion years of life, much is happening. Within that long period of time, microbes gain the ability to live independently, some work together, others metabolize hydrogen sulfide, and still others learn to photosynthesize—to turn sunlight, water, and carbon dioxide into sugars and oxygen—and these last group of microbes would forever change the planet.[5] The invention of photosynthesis potentially meant extinction for countless living things for which oxygen was toxic or a dangerous mutagen.[6] Oxygen paired up (O_2) is great—we need it, and so do a multitude of other organisms that depend on it to survive. But alone, oxygen is a "free radical." That's why your favorite sports beverage label is always talking about "antioxidants." Free oxygen is a 'radical' because it binds to random stuff, like your DNA, and causes mutations.[7] And even though some, perhaps most, of those mutations were either lethal or inconsequential, many others were beneficial. After the "Great Oxygenation Event" came the great diversification of "eukaryotic" life, multicellular organisms that were at first just more instances of unicellular life with a separate nucleus with chromosomes of DNA.[8] Eventually, these eukaryotes would incorporate mitochondria, which needed lots of atmospheric oxygen (O_2, which makes up about 20 percent of our air today) to make energy. And, eventually, these supercharged eukaryotes would also become multicellular; plants, fungi, and animals became multicellular several times within those lineages.[9]

The multiple, independent origins of multicellular life remains a real mystery.[10] Some people point to a global ice age, a "snowball Earth" period about 650 million years ago, that may have contributed to the multiple origins of multicellular life, as if these evolving organisms had to huddle together and learn to cooperate to survive the cold.[11]

I often think that alien life, which is almost certainly out there somewhere among the trillions of planets, moons and stars, will likely turn out to be as boring as life on Earth was for its first few billions of years. But, who knows, if we're lucky, maybe aliens will turn out to be more like E.T. or Chewbacca. And while on the subject of science fiction, I always liked that in H. G. Wells's *The War of the Worlds*, it isn't machine guns and tanks that defeat the invading Martians but microbes, whom we've evolved with for ages, that take down the sterile uninoculated extraterrestrials.

So maybe microbes aren't all that dull, but let's move to creatures we're more familiar with. About 540 million years ago we have an explosion of life: the "Cambrian Explosion." Multicellular life gets a backbone, and with it we get the first fishes, from which all other animals with a backbone spring, all with the same basic body plan.[12] We also get an amazing variety of organisms, including some that don't survive for very long and that we have a hard time linking to modern life forms.[13] Evolution was experimenting with different body types, and the ones we see around us today are the ones that survived from that time.

While microbes keep evolving, so does multicellular life, but still largely confined to water; it is really only about 500 million years ago that plants, fungi, invertebrates, and eventually fishes start to colonize solid land.[14] (New fossil evidence suggests unicellular fungi got to land first as early as a billion years ago,[15] and molecular evidence points to plants and invertebrates coming soon after.[16] From our fishy ancestors, we get all the land vertebrates you can think of today, but we don't get mammals until about 200 million years ago, some are as big as the dinosaurs, but most are much smaller, the size of rats or bats, just like most mammals today.[17]* From these rat-sized creatures, we eventually get the lineage that gives rise to humans. But when I say "humans," I don't just mean *Homo sapiens*—until pretty recently, there were a dozen or so species we could call "human" or at least "homininans" (members of the Hominina: the branch on the Tree of Life that is the sister group to chimpanzees and includes us, *Homo sapiens*, and our closest extinct relatives).[18] But that's just our part of the Tree of Life, where we're just a "tiny emerging twig," in the larger scheme of things.

Today we have flowering plants, ferns and fungi, paramaecia and bacteria, ducks and whales, and all manner of other living things evolving all around us. And, although it's not surprising that we humans tend to focus on ourselves, we could stand to be more mindful of the millions of other living species on our planet. Instead, we're replacing the rich and wide diversity of life around us with the relatively few kinds of life that we prefer or that "benefit" us.[19]

---

* I say that mostly to annoy my mammologist friends, but the average person thinks mammals are way more diverse than they are—it's not all platypus and pandas, kids; 75 percent of mammal species are rodents or bats, and, by weight (and roughly by number), wild mammals account for only 4 percent of all the biomass of mammals on Earth (humans and livestock making up the rest).

Most mammals (by weight) are now livestock, far outweighing wild mammals. And, in fact, humans plus livestock far outweigh all other terrestrial vertebrates combined.[20] We are replacing natural selection and neutral evolution with rapid artificial selection. And we're not just causing extinctions—we are fundamentally changing how evolution works on our planet.

IV

WHY UNDERSTANDING EVOLUTION MATTERS

EVOLUTION IN THE ANTHROPOCENE

I have a French Bulldog named Bessie (figure 14). She's a rescue dog because I'm a fan of places that save dogs, rather than those that breed them. Although I would like to believe in a world where there are bands of French Bulldogs freely roving and rolling around in the Pyrenees, they are, in fact, not found in the wild. Like all dogs, French Bulldogs (or "Frenchies," as they're affectionately called) are descendants of wolves. Bessie is cute and sweet—but her short, snorty, bug-eyed features are an affront to the regal dignity of her wolf ancestors. Her truncated snout is the opposite of what has been happening for millions of years of canid evolution[1]—something I think about when she sleeps with her head in my armpit and snores away like a tiny goblin. As with all pure breeds, breeders mixed and matched different pairs of individuals, selecting the mates that would produce offspring with the same traits as their parents—or with "improved" traits. (There is a good reason Darwin began *On the Origin of Species* with examples of domesticated animals.)

In the case of Frenchies, they were bred to have short muzzles, strong, stout bodies, and a temperament suited for companionship.[2] The unintended consequences of having

Figure 14 My French Bulldog Bessie.

their cute, short snouts and large heads, however, are that they have trouble breathing, can't cool down well (because of short nasal passages), and have hips too narrow to give birth naturally: their babies must be born from C-sections. In fact, their hips are so narrow that the male can't even mount a female, so females must be artificially inseminated. Is that evolution? No, not in the natural sense—but it is selection. In this case—and in every breed that humans have molded through artificial selection, be it bulldog, corn, or cow—it is we, not nature, doing the selecting. Nature (natural selection) also selects, of course; those organisms which are best adapted for an environment can survive and have offspring, while those unable to survive eventually die off with no descendants. Nature, also changes, from year to year and from season to season, and has no preference for particular traits, but we humans do have a preference when we are doing the selecting, which is why natural selection is a far slower process than artificial selection. Of course, artificial selection also takes a very long time: you have to wait for the organisms to reproduce over many generations. It takes many tries to get the few individuals with the traits you desire. To speed selection up with animals, mating siblings with similar traits is often part of the process, which leads to lots of early deaths and unwanted mutations (the likelihood of dangerous and harmful recessive traits coming together goes way up when you mate close relatives with each other).[3] But as if that forced incest wasn't bad enough, we humans are now doing something far worse: we are not just exploiting the laws of nature anymore; we are rewriting the code.

A DNA- or gene-editing tool called "CRISPR-Cas" (sometimes just "CRISPR," short for "clustered regularly interspaced short palindromic repeats") was recently invented by Jennifer Doudna and Emmanuelle Charpentier (together they were awarded the 2020 Nobel Prize in Chemistry for this invention), and it has revolutionized the world of geneticists and biotechnologists. CRISPR-Cas is a combination of a DNA-cutting enzyme (usually Cas9 or Cas12) and special repeated DNA sequences (CRISPR) that is the naturally occurring genetic mechanism some bacteria use to fight off infectious viruses. Essentially, CRISPR-Cas can cut out strands of infectious DNA and replace it with a similar but harmless sequence that helps the bacteria "remember" (so it will be ready to defend itself against that virus in the future), think of it as the bacteria's immune system.[4] This natural mechanism can be co-opted by scientists as a gene-editing tool (often called a "gene-editing typewriter" or, more aptly, a "gene-editing word processor") to do something similar: we can use CRISPR to turn off genes or modify their function to study how changes we make in the genotype (DNA sequences) of an organism can affect the phenotype (the body) of that organism. CRISPR lets a scientist, or a nineteen-year-old undergraduate lab tech, change the DNA code of an organism by editing or deleting its DNA sequences, much like a word processor lets us edit or delete words. Instead of waiting hundreds of generations for desirable mutations to pop up, we can use CRISPR to speed up artificial selection by manipulating an organism's genome directly. And using another genetic tool called "gene drives" with CRISPR-Cas, we can make any particular trait act like the only version (or "allele") of a gene and pass down that version to 100 percent of individuals in a population of organisms, generation after generation.[5] What do you think? Will we use these new tools for good or for evil? Both, of course. (To their credit, Doudna and Charpentier have been doing an excellent job reflecting on both the power and the ethics of using their new invention.[6]) To be fair, gene editing has been around for a while, but the ease and utility of CRISPR-Cas is unprecedented.

We have already started using CRISPR commercially for profit. There are CRISPR-edited fruit at the supermarket (such as "anti-browning apples") that age more slowly than normal "unedited" fruit, thus improving shelf life and cutting down on waste, and, of course, for making more money.[7] Although "genetically modified organisms" (GMOs) are not new, CRISPR is going to ramp up the speed with which we see them at the market. If used the right way, this technology has the potential to solve or at least alleviate food insecurity, although, of course, CRISPR may also be used in the wrong way, to grotesquely produce new animals, for example.* On the other hand, you could argue that artificial selection has already led to producing new animals, such as broiler chickens that are so fat they can barely walk and that are almost unrecognizable as descendants of their noble ancestors, the wild Asian "red junglefowl" (*Gallus gallus*).[8]

Once genomic engineering with CRISPR becomes more mainstream, perhaps a new type of bioterrorist will produce a mosquito that bites many people before laying eggs (it is usually just one blood meal before laying eggs for mosquitoes that bite humans). That modification would make mosquitoes even more dangerous disease vectors by spreading their blood-borne diseases, such as malaria, dengue fever, West Nile virus, chikungunya, and yellow fever, more rapidly and efficiently. We are already releasing "CRISPR-ed" mosquitoes (modified to make them less dangerous as a disease vector) in places where the diseases they carry are wreaking havoc.[9] Those releases are actually done by scientists and authorized by politicians who are being cautious; although where they carry out this work—often islands with large indigenous communities—is ethically troubling, to say the least.[10] Using gene drives, we can make mosquitoes that are resistant to these diseases and have that trait spread to all members of a population in the wild. But our regulations and ethics are right to caution us from releasing those "CRISPR-ed" individuals into the wild on a large scale. The bottom line is that

---

* For a better sense of how far we are going with CRISPR, check out the Netflix documentary series *Unnatural Selection* or Walter Isaacson's book *The Code Breaker*.

we can't be sure that the CRISPR or gene-drive mechanism producing resistant mosquitoes won't "escape" our artificially selected target organisms and jump to nontarget ones via viruses or some other mechanism. We don't want our release of resistant mosquitos to wind up producing sterile bees, sterile spiders—much less sterile people.

What if, intent on removing the organisms (besides us) most responsible for bringing about global warming, an eco-terrorist decided to use CRISPR on an existing disease virus to make one that would kill off all cattle? Or what if someone were to use CRISPR to modify a disease-causing microbe that already exists in animals, like the microbe that causes avian malaria, so it would jump to humans—just as the coronavirus jumped to humans from pangolins or bats in 2019?[11] By the way, because we understand how viruses can be altered in the lab, we can disprove the conspiracy theory that the coronavirus that causes COVID-19 (SARS-CoV-2) was an altered virus made in a lab. None of the genetic signatures that we would expect from an experimentally altered virus can be seen in the DNA sequences of SARS-CoV-2; instead, the COVID-19 virus is very similar to the SARS virus that already existed in wild animals before it mutated—on its own—for animal-to-human and human-to-human transmission late in 2019.[12]

Imagine a future where new species are produced in labs and released in the same way exotic plants and animals are released to this day. If you go to Florida or Singapore, you can see non-native plants and animals almost as easily as you can see natives.[13] In the United States, you need look no further than your backyard or garden—where such non-native species may include house sparrows, starlings, fire ants, and coconut palms, while, in Asian countries, they may include lantana and water hyacinth, African snails and tilapia.[14] Some of our most pervasive "wildlife" species are nonnative "invasives," although, in the future, it may be CRISPR-ed species that are the most abundant. I'm an optimist about many things, but not about our ability to always use CRISPR wisely. People are already bending or evading regulations in place to keep CRISPR use in check, which has resulted in the "CRISPR babies" recently born in China and the threat by a Russian doctor to "cure" some forms of deafness by "CRISPR-ing" the embryos of deaf couples.[15]* On the potentially bright side of this new technology, CRISPR might help rid us of harmful invasive species by spreading deadly mutations specific to that species or by making plastic-eating bacteria more efficient at consuming, degrading, and digesting all kinds of plastic waste. Indeed, by helping save the lives of those with sickle cell anemia and other diseases with improved gene therapies, CRISPR is already having a hugely positive impact.[16]

Before humans, there was natural selection, and it worked for billions of years. It still does, but its preeminent position is being challenged.

Humans performed artificial selection first by using selective breeding and now by CRISPR; with ever-increasing speed

---

* For many, and not only those in the deaf community, deafness is not an "illness" that needs to be "cured." Are we going to do away with conditions such as Down syndrome and autism by "curing" them into extinction? For some, that's a call for the extinction of communities that borders on eugenics. To look at the historical context of these issues as they may relate to the use of CRISPR-Cas in humans, see Carl Zimmer's fantastic book *She Has Her Mother's Laugh*.

and skill, we are becoming the real "Intelligent Designers" of the world around us.[17] We're not just modifying the organisms that we exploit like cattle and food crops but we're also reshaping the planet in ways that even the organisms we're not interested in modifying are affected secondhand. And the secondhand effects on other organisms from what we're doing are not just limited to our artificial selection. Take trophy hunting and fishing, for instance, where we bag or reel in so many of the biggest and most decorated (horned, tusked) or most beautiful (striped, brightly colored) individuals of game species that the size and appearance of those species, and even the role they play in their ecosystems are fundamentally changed. Removing these trophy-worthy individuals from the population of a species means that the genes the individuals carry for those traits are taken out of the gene pool and leaves smaller, less decorated, and less well-adapted individuals to breed and pass on the genes for those less-impressive traits to their descendants. Elephant tusks are not as large and heavy as they once were before humans started hunting elephants (and tusk-less elephant populations have recently evolved);[18] bluefin tunas aren't nearly as massive as they once were before they became prized by wealthy foodies.[19] Not to mention the hundreds of species we have hunted completely to extinction: passenger pigeons (once the most numerous bird on Earth), the dodo, the Tasmanian tiger—these are just the most charismatic or notable examples.[20] Some estimates are that, relative to the normal background extinction rate (remember, some extinctions are just a part of evolution), humans are causing hundreds, if not thousands, of additional species to go extinct every year (many of these are invertebrates, amphibians, and island species that don't get much press).[21] Most living things that have gone extinct during the existence of human beings are being lost because of habitat destruction, which includes pollution and removal or reshaping of landscapes.[22] Some animal species normally active in the daytime have even shifted to a more nocturnal existence when they live near us.[23] Next time you have the window seat on a flight, keep an eye out for how much forest you see versus something modified for us humans. You will likely see a tapestry of destruction. The scope and profound consequences of our reshaping both life on our planet and the planet itself during our species' time on Earth are why many call the most recent part of our current geological epoch the "Anthropocene."[24] We humans are creating the same level of impact on the Earth as a meteor strike or an ice age.

So, what to do about our ever-worsening situation? That is for each of us to decide personally. I'm not here to tell you to become a vegan or to stop using plastic straws when the solutions to our massive environmental problems need to be undertaken at a governmental level and on an industrial scale. Personal choices and actions help, but unchecked industrial pollution and indiscriminate destruction of rainforests and coral reefs need action beyond what you and I can do on our own. It is interesting to me that, even in the thick of this coronavirus pandemic when so much of humanity was sheltering in place, we still didn't make a big enough dent in pollution to hit positive climate change benchmarks.[25] I'm certainly not saying we shouldn't put in our individual efforts—it's just that we also need to have change at a community and governmental level. I fully admit that I don't know how to keep an economy going or how to run a business and that not all the answers to our most pressing environmental problems can come from science, but scientists can warn us about the future consequences of our actions. And that's the real purpose of this chapter: to explain not just how we are reshaping the planet and the organisms that live upon it, but also how we are changing the ways evolution, itself, works. We are altering the future of life on Earth and rewriting the evolutionary rules of the past with no clear understanding of the implications of what we're doing in the present.

NATURAL HISTORY

We are on the verge of changing the rules of nature with tools like CRISPR without a clear understanding of our natural world, let alone an accurate and comprehensive inventory of all its inhabitants. When they think of evolution research, many folks think of genetics labs with scientists in white lab coats at work in sterile modern facilities. I think of someone wading through a swamp waving a thin net in the air collecting insects, someone else looking at the behavior of bowerbirds through binoculars, and another someone slipping into a cave to collect some previously undiscovered species. The study of evolution still means a lot of boots-on-the-ground discovery and what is often called "basic science research" (the kind of research that is fundamental for the advancement of scientific knowledge, but seldom directed toward some economic gain).[1]* The products of all those

---

* I am a former program director at the US National Science Foundation, one of the world's greatest funders of basic science research. I loved that institution's foundational charter—Vannevar Bush's "Science, the Endless Frontier," which laid out the importance of basic science. People used to call this fundamental basic research "true science," and it was seen to be in poor taste to do any applied "money-earning"

boots on the ground are natural history museums, herbariums, genetic resource repositories, and other research collections of plants, animals, fungi, and microbes. The DNA and bodies of organisms we natural history researchers use to study life on Earth were collected by biologists across the globe, and these collections represent our reference libraries for Earth's biodiversity, each species a book, each specimen its own special edition.[2]

Some people focus on the potential environmental impact natural history research might make, research that includes "taking samples"—that is, humanely sacrificing individual organisms to bring them back to the lab for study and to preserve them in perpetuity. You can't save what you don't know. These samples serve as reference "vouchers" that help verify whether a specimen has been correctly identified in terms of its taxonomy—that is, given the correct scientific name, so we know which species we are studying. And they also serve as evidence of the source material along with associated data that can be important for scientific repeatability (such as determining whether a specific genetic sequence really is from that individual specimen or was the sequence or specimen mislabeled?[3] Because the samples themselves can easily be misidentified, and often are (taxonomy being a dying art practiced by relatively few, especially for less charismatic groups like bacteria),[4] it is imperative that

---

research as a scientist. That elitism kept Thomas Edison out of the US National Academy of Science for some time, and why neither he nor Nikola Tesla ever won the Nobel Prize. That view has flipped—the trend now is to look at natural history and other "basic science" as cheap and "basic," with much of the funding and attention going to the "applied" sciences like medicine or bioengineering.

researchers have a reliably identified voucher to use in their identifications.

Far from being heartless "collectors," natural history researchers take samples because we care about the larger populations of the organisms we are studying: "We sacrifice the few to save the many," as I like to say. We are doing what Aristotle and Darwin did: observing and describing what we see in nature. But sometimes we need to see what makes an organism tick, and that may include dissecting it or just studying its physiology up close. The species of many organisms can't be positively identified working from just photographs or videos; their bodies contain valuable data. Or, as we museum folks often say, "A specimen isn't dead if it has data." We preserve the bodies of organisms because all the questions about them that we can ask today and in the future require that we do.

There certainly are "collectors," who want bodies to show off as trophies, no matter how rare or endangered the species may be. Perhaps the worst of these was Lionel Walter Rothschild, whose early twentieth-century network of amateur collectors engaged in unchecked killing and caused environmental disasters for the sake of Rothchild's private collections. Although he may have done this in the name of "exploration," he certainly did not act on behalf of conservation.[5] The work of modern natural history researchers is nothing like the selfish practices of Rothschild and similar collectors, but many don't see the difference. Natural history collections are so poorly understood these days that some people have stolen important and historic samples from them while justifying their crimes with the nonchalant attitude of someone stealing a library book. And they are punished about as harshly as jaywalkers by naive justice systems (read *The Feather Thief*, for instance[6]). Unlike a given library book, however, each sample in a natural history collection is unique and tells a different story about life on this planet.

So why can't natural history scientists just keep one specimen of each species? Because we need to understand variation in all different kinds of individuals of a species: males, females, juveniles, breeding adults, and so on. Go back to the earlier sections of this book to remember how important understanding variation is to studying evolution. Someone might ask, "Are all the little toads in this ravine the same species?" We could just take a picture and a DNA sample from each toad and return them to their habitat, but it isn't always so simple. Sometimes the variation is much subtler than we expect. We don't need all the toads to complete the story—besides, that would clearly be unethical—but one sample specimen likely won't answer all our questions either. No matter the question, the sampling protocol needs to be worked out and approved in advance for all the stakeholders involved (indigenous or other landowners, scientific or government agencies, or some combination of these and any other stakeholders).

I am studying a group of bioluminescent animals called "ponyfishes" (the family Leiognathidae), whose males flash with light displays to attract females. Across several species, the females look almost identical to one another but the males don't: the translucent patch on their bodies (where they make their sexy courtship light displays) is shaped slightly differently across species—sometimes it's triangular, sometimes rectangular, and sometimes half-moon shaped.[7] By dissecting or by imaging with CT or MRI scans, we can find bigger internal differences between the fishes. Each male ponyfish is home to glowing bacteria in a "light organ" around their esophagus and has mirrorlike skin that helps direct the light inside its body cavity to an area on the translucent surface of its body.[8] I don't feel bad about the jars with hundreds of ponyfish specimens in the fish collection I curate because almost all the specimens are from fish markets where I collected their already-dead bodies. They are so abundant that I'm usually not even asked to pay for them since they're often found in discarded garbage heaps of unwanted "bycatch." In fact, many of the dead ponyfishes I sample (along with lots of other "low-quality," high abundance marine species) are usually on their way to being ground up for use as fertilizer, or as cheap protein meal for cows and chickens, or even as cat food.[9] These ponyfishes are collected in the millions every day throughout Asia, Australia, the Middle East and Africa. They are sometimes picked up by "supertrawlers" and other poorly regulated fishing vessels. I've described four new ponyfish species to science from these "garbage heaps."

Many of the birds in our museum collections are from bird strikes on building windows or oil platforms, where birds die in the billions worldwide every year, not to mention the billions more killed by pet and feral cats.[10] By taking bird strike samples and labeling the specimens as evidence in our collections we know where and how they died, we can often help document some of the myriad ways humans are driving species to extinction, and that information can help lead to new window designs and new oil platform practices to help prevent bird strike deaths in the future.

In freely acknowledging that we also sacrifice animals when we take samples, we need to stress that the impact of natural history sampling is minuscule compared to the wanton death we humans bring upon these species through our introduction of invasive species, habitat loss, overfishing, and any of the myriad other destructive things we do. For instance, it is estimated that the recorded total annual catch of the fishing industry amounts to the deaths of some 5 million individual fish per *minute*,[11] which is more than ten times the number of specimens in the fish collection I curate from more than *fifty years'* worth of careful, selective collecting. Unlike those who engage in (or are responsible for) wanton acts of destruction, the vast majority of natural history scientists take sample specimens only to use in our scientific studies of species and evolution or to assist the future work of other scientists from a wide range of different disciplines.[12] A typical natural history museum lends hundreds of specimens and DNA samples a year to researchers studying everything from climate change to new medicines (half of medicinal pharmaceuticals are derived from nature[13]). We could more rapidly combat the outbreak of new diseases like coronaviruses (SARS-CoV-2) and Ebola if we had specimen vouchers in natural history collections that could help us detect the sources of these diseases in the wild.[14]

An ongoing project at the museum where I work involves seeing whether the pollen found on mammal specimens can identify where drug traffickers are hiding.[15] More commonly, collecting and storing specimens is a critical part of conservation biology, helping us understand how communities of organisms and the organisms themselves are changing over time.[16] When we sacrifice animals in collecting specimens, we kill as humanely as possible, as dictated by the strict rules of institutional animal care and use committees, and in compliance with the laws or regulations of the federal, state, local, tribal, or foreign agencies or governments having jurisdiction in the areas where we are collecting and from which we must obtain permits and permissions. That said, there is the problematic issue of "helicopter science," where researchers from Western countries swoop in to collect their data but share little in return with their local partners from less privileged countries. And, truth be told, at least some natural history researchers have been part of this problem. All the more reason why helping build scientific infrastructure in the places we work should be part not only of our natural history mission, but also of the mission of modern science itself.[17]

But why are we still doing natural history in the twenty-first century? In part, we're doing it to study—and even improve—the Tree of Life. We use the bodies and the DNA samples from the specimens we collect to figure out which species is related to which other species and to better understand evolution. In fact, before the study of evolution, when many people didn't think extinction was even possible, there was far more indiscriminate collecting, just for the sake of acquiring trophies for the rich guy's "Wunderkammer" (or "Cabinet of Curiosities"). After Darwin, we gained a deeper appreciation of what these collected specimens could tell us about the form and function of their respective species and about our own evolutionary history: every individual specimen gained in value, the importance of variation and the connection between the species of these specimens was revealed to us through our study of evolution—and collecting indiscriminately for collecting's sake became no longer sensible.

The science of natural history also helps us to add branches to the Tree of Life by discovering and describing new species. In the course of my career in natural history, I've described fifteen species new to science, and new species (both living and extinct) are discovered and described every day by researchers across the planet (about 17,000 new species a year; and not just boring ones either—we get a new dinosaur species about every week).[18] With knowledge from the Tree of Life, we can change the scientific names of species to better reflect their relationships (taxonomy)* with one another or to learn more about the evolution of certain traits like bioluminescence, which appeared independently dozens of times in evolutionary history. Or we can try to figure out how we can distinguish between major lineages and come to better appreciate that no matter how different a peacock looks from a hen-of-the-woods mushroom that, yes, we all share a single common ancestor, we all came from that ancestor who lived nearly four billion years ago. What an amazing thing it is that all of us living things are related to one another, and how wonderful it is that we can study our history and answer the question "Where are we from?" by pointing to the Tree of Life and exclaiming, "There!"

---

* If you don't think taxonomy matters, tell me the difference between "coronavirus," "COVID-19," and "SARS-CoV-2." The first is a class of viruses, the second, the disease caused by a specific virus in that class, and the third, the actual scientific name of that specific virus. The difference between them may not seem significant to you, but knowing it helps scientists be precise in keeping track of the information related to the three distinctly different subjects these three terms refer to.

OUR GENEALOGY AND ANCESTRY

Why should you care about where you belong on the Tree of Life? Because not only does it tell you *where* you came from, but also *who* you are. As humans, we tend to focus on the differences between us, no matter how small those differences might be, but we are all more similar to one another than many of us are willing to believe. As more and more people take ancestry tests, sending their money and saliva to 23andMe and other genealogy testing centers, we need to be educated on what the results of those tests actually mean scientifically, and we all have to decide, together, what they mean socially. Many of us are coming to learn that science describes gender, race, sex, and sexuality as attributes on a spectrum, but for much of our modern lives we've forced these attributes into a few arbitrary, usually binary, categories—and nature loathes a binary. Why look at the rainbow that is humanity in just black and white? If instead you look at that rainbow carefully you would celebrate the fact that all of that colorful diversity comes from the same little drops of water, each with just a little different shine. We humans focus on our differences because we share so many similarities. To the cosmos, we are all still just "star stuff," to quote Carl Sagan; all of us life forms are made up of the same chemical elements—hydrogen formed during the Big Bang and elements like oxygen, carbon, and nitrogen forged in the hearts of stars.[1] It's what we do with our star stuff that makes us special.

The more you learn about evolution and about "where you are from," the more you can see our similarities. An alien might be surprised how we focus on the small stuff instead of the star stuff. Twins can check off different boxes for race based on how they look and perhaps based on how other people see them. There are just a few genes that determine eye color and skin pigment—is that what determines your "race'? What does it mean when one sibling has dark skin and is treated as "Black" and the other fair skin and is treated as "white," as sometimes happens with mixed-race children?[2] I love the poem "Genetics" in Jacqueline Woodson's book *Brown Girl Dreaming*,[3] which talks about how strangers didn't believe her "pale as dust" brother was actually related to her darker-skinned family until they all smiled, revealing the same gap between their two front teeth. Do DNA test results show your "race" or family history as easily as those shared gap-tooth smiles? Not always. You can have a Native American ancestor who doesn't show up at all in your DNA test results but does in your sister's results. You might be African American but have some significant proportion of Neanderthal DNA that is usually seen at that level only in people of European descent. There are hard truths in your DNA, but there are also mysteries—and even miracles. Your genome is Pandora's Box, Aladdin's Lamp, and Alice's Looking Glass; understanding what your genome, can and can't tell you is part of understanding your origins.

I think we should separate "ancestry"—the people who are your actual relatives and ancestors that are in your family tree—from your "genealogy," which I think should only refer to what your DNA, within *its* limits, can and can't tell you about your ancestry. You probably have close to zero DNA markers left from your great-great-great-great-great grandfather, so he may not show up in your genealogy, but he's still part of your ancestry as a dead ancestor in your family tree.[4]* The reason for the paucity of DNA from that long-dead ancestor is that the DNA remaining from every previous generation is essentially halved with each passing generation. You get about half of the DNA in your genome from your mother and half from your father, and your brothers and sisters get different halves (via meiosis and recombination); that means you can get about one-quarter of your DNA from each grandparent and one-eighth from each great-grandparent, one-sixteenth from any great-great-grandparent and so on and so forth. These are expected, or perfect, proportions; but it's not that simple; recombination is not that exact, and you can lose all of that one-sixteenth pretty easily.[5] (Go back to the "Mutants and Mutations" chapter for a refresher on recombination; I promised you a salad metaphor back there, and here we are.)

Imagine that your genome is a salad. You get half your salad from your mother and the other half from your father. Your sister would get a different half from her mother and a different half from her father (imagine if their salad bowls

---

* The evolutionary geneticist Graham Coop understands the strangely odd dynamics of DNA heredity in humans better than anyone I know. Check out Coop's excellent blog at https://gcbias.org/.

are never empty and maintain the same composition, just tossed around before being distributed). Your family might be a family of Italian salads, but three generations ago, you had a Greek salad ancestor. Your sister might end up with some feta in her salad from that ancestor, but you might not, simply because of the unique combinations of egg and sperm from which you received your DNA. So even though you have the same ancestors, your genealogy might not reflect your complete family history. So it is with DNA "genealogy" testing: it is limited by the mixing and loss of genetic markers with each generation. These markers are lost because each egg and sperm cell from your parents has just one set (half) of their chromosomes that is a unique recombined mix of the pairs of chromosomes in their other cells. The sperm and egg come together to give you the full set of chromosome pairs again, but that's the reason you are a mix of your parents and not clones. (Along with a few dozen unique mutations that are yours alone.)

It is rather remarkable then that people of European descent generally still show about 2 percent[6]* of their DNA coming from Neanderthals, the equivalent of one-fiftieth or about the amount of DNA you might expect from an ancestor from six generations ago. However, *Homo sapiens* haven't mated with *Homo neanderthalensis* in more than 40,000 years

---

* In reality, this "2 percent" is only from the 1 percent or so of total variable DNA that most testing agencies actually examine. You can be both "100 percent French" and "2 percent Neanderthal." What these DNA tests are doing is called "genotyping," looking at specific variable regions of your genome. To learn more about testing and ancestry, check out my 2020 TED-Ed video "What Can DNA Tests Really Tell Us about Our Ancestry?'

(or 1,600 generations).[7] So, there wasn't just one Neanderthal ancestor for these people, but many. People descended from ancestors in various regions of Africa also have evidence of hybridization with other extinct humans, as do people of Asian descent (although, notably, this doesn't always match up with our concepts and constructs of race—see below). All of us humans also have a lot more shared recent ancestors than you might expect, even considering when human population numbers were low in parts of our history due to disease or colonization. There weren't always a lot of choices, and for many—no choice at all. Especially when you had to have children with someone of your own race, religion, social standing, and so on. Which means that a large part of our complex backgrounds includes a lot more inbreeding than we'd probably care to admit.[8]

Our evolutionary history may include some puzzling facts, but our more recent ancestry can be pretty perplexing too, especially as DNA testing becomes commonplace. In 2018, US Senator Elizabeth Warren took us down a slippery slope when she had her DNA tested to "prove" she had some Native American ancestry.[9] Having Native American ancestry is not just a matter of genetics; it has an important cultural significance. Ancestral tribal membership is something that can neither be claimed nor denied on the basis of a simple DNA test.[10] Although the senator's test results did show she had at least one Native American ancestor, they did not show, nor did Warren ever claim to know, the tribal membership of that ancestor. What outraged many Native Americans was not so much that Warren claimed to share in their ancestral heritage (although she was wrong to do so without any tribal association), but that she took a DNA test to "prove" it.[11] The senator's real mistake, however, was to make DNA testing seem like a way people could scientifically prove their ethnic or cultural heritage—when it isn't. By law, it isn't either: you can't use the results of a DNA test to prove that you deserve minority status, for instance.[12] These genealogy tests are good, but they are looking for specific markers of ancestry, markers that might have been lost through the complicated workings of recombination.[13] Which again is why I think we should separate the meanings of "ancestry" and "genealogy," which are currently used as synonyms.

Remember too that we are all remarkably similar to one another in terms of our DNA (less than 1 percent difference between any two people),[14] so there isn't much that makes us unique. It's hard to find something from any ancestor who wasn't, or any ancestors who weren't, that different from you genetically. And it's especially hard to do so if you have neither your own full genome (as few of us do) nor the full genomes of your ancestors, whether distant or more recent. Commercially available ancestry tests are really looking at genetic markers or combinations of those markers that you share with people living *today* and in a particular region—people making up what are called "reference populations."[15] Those genetic "references" don't represent everyone in an area either, but just a few select people whose genotypes are used to represent "France" or "West Africa," for instance.[16] And the people living today in a given area are seldom the best representatives of the past: thus France or West Africa twelve generations ago almost certainly would have looked really different from what they look like today (thanks to war, racism, classism, immigration, and emigration, among other factors).

People interpret their "race" while learning about genetics in sometimes bizarre ways. There was a rally recently where white supremacists drank from jugs of milk, shirtless (of course), in cold weather.[17] What these men were actually trying to show was that they had the "lactase persistence" mutation that allows them to drink milk into adulthood (lactase being the enzyme that lets people digest the lactose in milk and other dairy foods), a trait they perceive as being part of their "racially superior" genetic constitution.[18] Many people across the globe lose the ability to digest milk when they leave the breastfeeding period of infancy—we call them "lactose intolerant," whereas some people from Northern Europe have a mutation that allows them to produce lactase into adulthood, perhaps an adaptive trait gained via natural selection by that population when dairy foods were more abundant than other foods and those who could digest milk had a better chance of surviving and reproducing.[19] But these white supremacist guys probably missed the part where their ancestors also endowed them with higher rates of gluten intolerance—the inability to digest foods made from wheat and other grains—which is maybe why they were drinking milk instead of beer.[20] The white supremacists misinterpreted the findings of earlier "lactase persistence" studies in popularized accounts[21] as proof their special mutation was evidence of their Aryan superiority (not so, but at least they were reading or hearing about the science involved).[22] West Africans also have high rates of lactase persistence, as do many other "nonwhite" people around the world.[23] Ironically, people from Southern Europe, including folks from the Caucasus Mountains, after which "Caucasians" are named, don't have that lactase persistence mutation quite as often.

These racist ralliers need to get their facts straight, including that "Aryan" was never a "race" even by most cultural standards. (By the way, this use of meat and dairy consumption in racist tactics isn't new.)[24]

There is no biological reality to what we call "races" in *Homo sapiens*—"race" in humans is a social construct.[25] We are all blended shades in a tub of mixed paint. Yes, you can still see some streaks of black, white, brown, and those of every color of human, but it's all made of the same stuff. Some "Caucasians" in the United States may never consider people from Mexico to be part of their own race. If they choose not to call them "Mexican," they might call them "Hispanic," "Latino," or "Latinx" by default, but of course many people from Mexico and the United States can both be descendants of the same European ancestors and could both be "Caucasians." "Hispanics" are just people from Spanish-speaking countries.[26] They do not constitute a biological race—people from Spanish-speaking countries can be Black, white, and every shade in between. "Latinx" (a term used more often in the United States than anywhere else) are just people from Latin America.[27] Is that a race? No. Not in terms of biology anyway, but that doesn't make "Latinx" and the other category terms we've discussed here completely useless either. We do need racial or ethnic categories—and even "social" categories, too—in the United States and other countries with similar histories. Why? One reason is to identify disparities in how people have been treated based on the perceived categories into which they've been put. For instance, African Americans were enslaved, then brutalized, and later left unprotected by the civil rights legislation enacted after the Civil War and in the 1960s, so maintaining some "social" categories, such as "historically underrepresented" or "continually underserved," would help the advance to equality for marginalized groups, whether social, political, or economic.[28]*

I am "Indian," but as a "racial identity" and not as a nationality, because both my parents are "Indian" and I was born in Montreal. How long does someone with Indian parents remain "Indian"? Are my children, who were born in the United States, and whose mother is not Indian, "half Indian"? Will my grandchildren be considered "Indian"? When do you become something else? Are people from Sri Lanka or Bangladesh "Indian"? Are all of us from these and other Asian countries just "Asian"? That's the box we check off for "race." All 4.5 billion of us just get one box? (In the United States, people from anywhere on the Indian subcontinent are most often labeled "Asian," as are Chinese Americans or Japanese Americans.)

Am I Canadian? I have a birth certificate that says so but I've lived in the United States most of my life. Am I American? Is there such a thing as "American" in terms of race? Some people with red hats sure think so. Is my wife from Montreal "Canadian" or "French-Canadian"? Is she "French"? "Non,"

---

* I can't recommend Angela Saini's book *Superior* enough for its scientific but highly readable take on racism in science and culture. And while on the subject of equality in science, I have to call out my own colleagues in evolutionary biology, very few of whom are Black, a lack of diversity best explained by systematic racism in our own and other academic fields. We academics also need to be more inclusive toward the indigenous communities we have failed, over and over again, particularly with respect to our genetic and genomic research. Unfortunately, evolutionary biology has a long history of being on the wrong side of racism in science.

she would say in French. What is race, ethnicity, nationality really?

Are people from Egypt "Africans"? How about South Africans or Kenyans? They all live in or come from Africa, yes, but their histories and those of their countries are vastly different. There is no race of people who are "African." Africa is a place with many "races," if you want to use that term. In fact, biologically, and in terms of evolution, we are all "Africans" as part of the same single human race and species, whose origins are unequivocally African because all our earliest human ancestors can be traced back to that continent.

Many people who identify as "African American" in the United States are the descendants of a stolen and enslaved people forcibly brought there.[29] Often, they don't even get to know whether, for instance, they have kin in Sudan or are descendants of the people from the Congo region or elsewhere. Their past was taken from them. They have a right to identify their ethnicity as "Black" or "African American" since they cannot always find evidence of their regional ancestry. Similar situations can be found in other parts of the world among other groups of people. (Consider how some Indians fought to remove the label of "untouchables" while others fought to find respect by recognizing their status as an historically oppressed class.)[30] There is no single answer for how people can be recognized in terms of race or social groups even biologically, it is often up to them how they interpret their identity; although, I'm definitely not saying it is a choice either (thinking of those folks who intentionally fabricate their identify as Black without having any Black relatives).

Many people think all Africans are Black and forget about the diverse people from Libya, Egypt, Algeria, and Morocco.

In the United States, people from these North African countries may get lumped together as "Middle Easterners," even though they actually come from North Africa. In Kuwait, which actually is in the Middle East, most of the people I saw working there were from Bangladesh, and were never treated as Kuwaitis no matter how long they had lived there. People who are actually from the Caucasus Mountains (as opposed to typical "Caucasian" Americans) include Armenians, are often dismissed as "looking Middle Eastern." Why is Barack Obama "Black"? Because of his outward appearance, how he saw himself and how people treated him, dictated it.[31] What about other mixed-race people—Derek Jeter, Keanu Reeves, Bob Marley, Bruno Mars, Kamala Harris? In the past, mixed-race people didn't have a choice—those in power decided what your race was by how they thought you looked or by the race of your forebears or ancestors, no matter how distant. In the United States, anyone with "one drop of Black blood" was considered legally Black—end of story, as late as the twentieth century.[32] However, just because "Black" and "white" don't turn out to be distinct biological groups (and they never were) doesn't mean we get to ignore disparities in how people are perceived and treated.[33]* To paraphrase Joseph L. Graves Jr. (the first African American in the United States to earn a PhD in evolutionary biology), "Just because races don't exist in an evolutionary sense doesn't mean there isn't racism."[34]

---

* Yes, I capitalize "Black" but not "white" because "Black" is considered an ethnicity, and "white" is not; it's just a skin color of many ethnicities. See also the latest version of the *AP Stylebook*.

SEX/GENDER/SEXUALITY

"Sex" is complicated, much more so in the expression of physical traits than of the physical act. It is perhaps even more complicated than "race," and, as we have done with human "races," we have historically pigeonholed the spectrum of sex into discrete categories that don't actually fit reality. Our contrived social abstractions often include treating "sex" as a synonym of "gender," and equating some forms of sexuality as "normal," relegating all others to "abnormal" choices or preferences. It's time our views evolved and incorporated science into our understanding of these very normal human traits in their full spectrum of variation. Like all human traits, sex, gender, and sexuality can be explained in the context of evolution, but I want to stress that you shouldn't need evolutionary explanations to accept people for who they are. Unfortunately, however, those who are less accepting often do ask why certain traits exist "if evolution is true." So here I will try to dispel those myths and explain why the human variation around the different spectrums of sex, gender, and sexuality don't contradict our scientific understanding of evolutionary biology.

When I would push my identical twin babies in their huge orange double-wide stroller through a crowd of gawkers, someone would eventually ask, "Twins?" "Yep, identical," I'd reply, and sometimes I'd then get, "Boy and a girl?" and I would shake my head, thinking, "Don't these folks know what 'identical' means?" But I was the one who had some learning to do. Only later would I learn that some identical twins, who have matching genomes, could still end up being different sexes, and, of course, different genders too, including being nonbinary or gender-fluid. One way is that during the formation of the sex cells (during meiosis), either the egg or the sperm might get two sex chromosomes rather than one, and then one of the developing twin embryos might end up with an odd number of chromosomes, so you might get an "XXY" male ("Klinefelter syndrome") born alongside an "XX" female twin.[1] Or, you might end up with a single "X" female ("Turner syndrome") alongside an "XY" male twin.[2]

There are actually dozens of ways your anatomical sex (whether you have a penis or a vagina) may not match your chromosomal sex (presuming XX female or XY male).[3] One way is "androgen insensitivity syndrome," where you do not express the male-associated genes that are on the Y chromosome.[4] The genes that control "maleness" are actually remarkably few, and they can move around, sometimes even onto the X chromosome. How "male" or "female" (and everything in between) you may be physically depends largely on how your hormones signal from relatively few genes, like one particularly fickle gene called "SRY" (which is short for what it codes for, the "sex-determining region Y protein"). SRY is perhaps the most important gene for male sex-character development, but it doesn't always stay on the male-associated Y chromosome.[5] If, as a man, you still link your "manliness" to your Y chromosome, you might also be shocked to learn how much that chromosome degrades as you age, and how it can be lost altogether in most of the cells of many older men.[6]

The conditions described above are far more common than previously thought. Imagine if you discover you are an XX male, or XY female? I bet you don't know which sex chromosomes you have (your "chromosomal sex"). You just assume you have sex chromosomes that match what you have in your pants; but you might not. So be kind to those who may express their sex differently than you expect; you, yourself, might have chromosomes and genes that tell a different story from the one you tell. There are also many influences that can impact how someone might be "intersex" (not being male or female but having a combination of sex related biological traits) to varying degrees.[7] Research puts the proportion of intersex people at around 1.7 percent of the US population, or roughly 5.6 million Americans.[8] There are complex interactions between genes, hormones, and environment that determine how we express our sex. And that goes for nonhuman species too, environment plays a crucial role in determining sex in some vertebrate species (e.g., incubation temperature determines sex in many reptiles).[9] We still have a lot to learn about sex and how it is expressed, even in humans.

Along these lines, we also have a labeling issue with a term that is specific to humans: "gender." Sex is one thing; gender is quite another. Gender is an expression of social and sexual identity unique to human cultures (and one that varies from culture to culture). You could say that gender is in your head, while sex is in your pants. In our culture, most people with male sex organs identify as men, but some identify as women. Similarly, some men have female sex organs, but most people with female sex characteristics identify as women; you might also identify outside of the male/female binary categories that are common but not fixed in many cultures. Your gender may also change over your lifetime, as it is a fluid trait for many people. The gender binary stereotype is often linked to sex and sexuality, but neither of these is binary; both are on spectrums across the Tree of Life.[10] Genders don't exist as a trait on the Tree of Life; we made them up for ourselves.

Although genders don't exist in the animal kingdom, that doesn't mean there is no variation in behaviors related to expression of a given sex. There are species of animals where the males behave or appear to be more like the females to avoid competition.[11] These "sneakers" use the advantage of appearing more like a female to mate with females when the bigger, more aggressive males are otherwise occupied posturing, threatening, or attacking one another. Although some would call what these "sneakers" do "deception" or "mimicry," this does not explain why a female would prefer a "feminized" male to mate with.[12] She isn't being fooled.

Sex can also change over the lifetime of many species. Have you seen *Finding Nemo*? If it wasn't a children's movie, what would really happen when Nemo's mother and siblings get eaten? Well, in nature, a little male clownfish like Nemo or his father Marlin would eventually become *female*—and a big, dominant female at that. Why? Clownfishes are sequential hermaphrodites; they are all born male and change into females when the opportunity arises, like when the big, dominant mama clownfish gets eaten.[13] (Real life doesn't always make for the best children's movie.) But, even in humans, your physical sex may change over time. In the Dominican Republic, there is a small community of children called "Güevedoces," who appear female at birth until about the age of twelve, when they develop penises.[14] This early but passing testosterone deficiency that causes this change is common in that community, but occurs elsewhere, too.

Again, sex ≠ gender. Gender is more fluid than sex, and many people feel like they were "born in the wrong body." There is a lovely book, *Becoming Nicole*, about a real family dealing with one of their identical twin boys who from a very early age showed an intense desire to be a girl.[15] This kind of true story is still unfamiliar to many of us only because we (at least we in the West) have just begun to acknowledge and accept that some people may have similar feelings about their assigned gender. Other cultures have been much more accepting of recognizing fluid gender expression, especially some parts of Southeast Asia, as well as some Native American communities.[16] Both twins in the *Becoming Nicole* story were assigned male at birth, but only one identified as female, Nicole, from the age of two. You may be wondering why, if gender identity is presumed to be at least in part genetic, did these identical twins, with the same genome, not express the same gender? Perhaps there is no genetic link to how gender is expressed, which is why it may be fluid over the lifetime of an individual. One other possible explanation is related to how genes are influenced by the environment ("epigenetics"). Identical twins aren't exactly identical because of epigenetic factors, such as the way certain molecules interact with DNA (usually termed "methylation"); differences in identicals can also be caused by less subtle environmental factors.[17] An identical twin raised alone in a cave with poor nutrition will look very different from one who grows up in a typical middle-class American family. Thus there may be both environmental and/or genetic factors that affect your sexuality and how you express your gender identity.[18] Although we don't know what those environmental factors might be, it's unlikely they include something your parents did (so you can tell that one uncle it is okay to sing Lady Gaga to the baby or to eat pineapple while pregnant). The fact that identical twins can express different sexualities and gender identities when raised in what seems to be the very same environment means there can be some pretty subtle environmental factors at play, factors that may involve the interactions of multiple genes.[19] Even nontwin siblings raised together in the same environment may be affected by such factors, as the Wachowskis (who created the *Matrix* movie franchise) may have been: born as male siblings two years apart, they are now both trans women. In about 20 percent of cases where one identical twin is trans, both are.[20] The fact that identical twins more commonly share the same gender identity than nonidenticals does point to a possible hereditary genetic factor (although both kinds of twins often share the same or very similar living conditions).[21] What all this means is that gender, as we currently understand it, is harder to study and explain than either sex or sexuality. So what about human sexuality in the evolutionary context? I'm sometimes asked something like, "Why are there gay people, if so much of evolution is about reproduction and passing on your genes?"

Homosexuality has been documented in hundreds, if not thousands, of animal species.[22] It's not uncommon, unusual, or "unnatural." My favorite example of same-sex mating in animals that I've seen personally are the three male elephants I encountered at a Sri Lankan preserve. The park ranger lovingly called them the "gay boys." This miniherd was made up of a thirty-, a twenty-, and a ten-year-old elephant (I'm not sure what the age of consent is in elephants). Usually, male elephants are solitary, but these three were always together and often . . . "displaying their affections."

We can observe and have observed same-sex mating in sexually reproducing animals across the animal kingdom,[23] but gender is not so "straightforward" as sexuality in animals. An evolutionary biologist colleague, Jeremy Yoder,* once told me, "You're really not going to find a clear analogue to human transgender identity in the nonhuman world, because animals really only have gender in the sense that they have social behaviors and signals involved in mating." That is, it would be a lot easier to infer gender expression in animals if all you had to do is notice, for instance, that a male elephant is acting like a female elephant, but most of the time elephants are, well, just acting like elephants, and you can't tell them apart by their behavior. Anyway, there really is no need to impose a gender on animals that clearly don't have one. Which leads me to wonder: Do humans *need* genders? I'm not sure we gain anything from these restrictive labels. If elephants don't need them, maybe we don't either? Certainly, I can't think of why we need genders in an evolutionary context.

---

* Learn more about Jeremy Yoder from his website: https://jbyoder.org/.

As a teenager, I worked at the New York Aquarium in Brooklyn, where two of the male penguins would later live and be celebrated as a same-sex couple. Same-sex pairings among captive penguins are not all that uncommon, and same-sex penguins that pair up are often given abandoned eggs to incubate and chicks to raise. Although some are even given "gay weddings," we can't really "gender" them in a way that parallels human behavior—nor do we need to.[24] (On the other hand, the fact that penguins already look like they're wearing tuxedos does make their weddings pretty cute, and having these weddings doesn't hurt anyone, while serving to promote openness to and acceptance of the full spectrum of sexualities.) There is no reason we should expect only binary genders in any species where we see a spectrum of sexualities; so why should we do so in humans?

Of course, scientists are looking for genetic links to these traits to better understand them, although you can't explain everything through genetics (see our discussion of gender above). Genetic studies of homosexuality have not yielded a "gay gene," although there may be some weakly associated genetic markers that can potentially, but not definitively, identify same-sex sexual orientation. Many people are afraid these investigations will just lead to more and worse bigotry, or even to a future where homophobic parents "correct" these potential markers with gene-editing tools.[25] The weak connections discovered between several genetic markers and same-sex behavior[26] didn't stop the ill-intentioned from producing a "How Gay Are You App" based on those results; worse still, it is distributed by a country that has criminalized gay sex, Uganda.[27] Such an app and any others like it will only result in further stigmatizing the LGBTQIA+ community. The search for "gay genes" is part of a new field of genetic research, often called "Genome Wide Association Studies" (GWAS) in humans, but we would be far better served if these studies didn't lead to more stigmatizing and demoralizing. The intentions of these studies may be to produce greater social acceptance by showing that particular behaviors have at least some genetic basis and are not simply "choices" or "preferences," but their findings can readily be distorted to show just the opposite, sometimes with terrible consequences. Many of these GWAS papers cast a very wide net, and forget correlation doesn't mean causation, particularly given the narrowness of the populations examined. On top of that, many traits are "polygenic" or influenced by many genes, and many genes are "pleiotrophic," meaning that a single gene can have a small or large influence on many traits. The weak associations of many of the markers being examined in these GWAS examinations must be put in the context of the overall genetic architecture of our complicated genomes.

GWAS investigations have also been conducted to try to find genetic links correlating to income level[28] and educational achievement.[29] I understand the motivation for finding correlates in newly available genomic datasets for a variety of important social conditions and behaviors, but there is also a great need for scientists and science communicators who can clearly explain to the public the significance and proper context of such correlates, especially those found to be weakly associated. Even Aristotle divided his writings into the "exoteric," for explaining his ideas to the average person, and the "esoteric," reserved for his fellow philosophers and preeminent peripatetics.[30] We would do well to remember that, poverty was once thought to be a heritable trait, one that American eugenicists wanted to eliminate through forced sterilizations.[31]

What does all this have to do with evolution? After all, same-sex couples can't make their own babies and pass down their traits to the next generation. Some people ask what the natural-selection explanation for same-sex attraction might be. It maybe shows their bigotry that we don't have similar evolutionary discussions when talking about the thousands of people each year who kill themselves before having children because of depression, anxiety, or because of any number of heritable illnesses.[32] That's because there isn't the same kind of bigotry directed toward people having those conditions or illnesses as there is toward people who are lesbian, gay, bisexual, transgender, queer/questioning, intersex, or asexual. Of course, being attracted to the same sex isn't an illness, but there are people who suddenly become scholars of natural selection and invoke Darwin's name in vain when they unleash their genderist or heteronormative bigotry against the LGBTQIA+ community. Even raising the question of why same-sex sexual behavior exists in an evolutionary context assumes that "normal" sexual behavior is different-sex sexual behavior, and it reveals a certain ignorance about what "normal" sexuality actually is across the Tree of Life.[33] The truth is you don't need a *biological explanation* to treat people with respect. It isn't "wokeness," just kindness.

About 7.1 percent of people surveyed in the United States self-identify as gay, lesbian, bisexual, or trans, which amounts to about 24 million Americans.[34] Even that huge number may be an underestimate and might explain why evolutionary studies on human sexual orientation are generally difficult to conduct. Many people still live in fear of the prejudice they might face if they were to reveal this aspect of their lives.

When people ask "Why are there gay people?" in an evolutionary context, they are really asking how come their "gay genes" don't disappear if they aren't producing offspring. One hypothesis for the existence of same-sex sexual orientation is that some of the genes for this trait may be passed down by people who also have sex partners of the opposite-sex with whom they produce children.[35] There are also hypotheses that the genes for same-sex sexual orientation may be carried and passed down by relatives who are "straight" (of different-sex sexual orientation) and who carry these genes but do not express "alternative" sexualities themselves. And, in fact, there is evidence that the female relatives of gay men have more children than those without gay relatives, which could lead to the trait persisting, as these more fecund women pass down those "gay-ish" genes even if they themselves may be strictly heterosexual in their orientation and behavior.[36] In that scenario, the loci influencing same-sex sexual orientation would express themselves differently in these women (than in their gay relatives) and be passed from generation to generation by the straight (or at least the baby-making) relatives.[37] There are nonreproductive members in many animal populations, and they pass on their genes by helping close relatives reproduce. Genetically speaking, you are passing down more of your genes by helping a sibling produce more nieces and nephews (who share about a quarter of your DNA) than having just a few of your own children (who share half of your DNA). You can imagine this system leading to more help for these fecund mothers from relatives who do not have their own offspring; those nonreproductive "helpers" in these scenarios are conjectured to be predominately gay.[38] But most of these "gay helper" hypotheses and scenarios for humans are a little too simple and convenient in my opinion; "more science please!"

What we do know is that sex, sexuality, and gender do not track with one another as we once thought they did. Not only do people have a wide variety of sexual orientations in terms of who they're attracted to, but many people are also born intersex in a multitude of ways (which they may not even be aware of), and gender can be expressed differently even within the lifetime of a single individual. There is variation in humans, as there is in all species, and although many societies have created social norms including binary categories for gender, sexuality and sex, we know that those categories do not cast the net of human variation widely enough. We need to do better, we need to educate one another and use inclusive language, and we need to be informed by the latest science to do so.[39] We should also not let misinterpretation of that scientific information lead to policy decisions that restrict freedoms based on behaviors we know are natural and variable. Perhaps we can even benefit from our struggles to define things like "gender" and "sexual orientation" to better understand how mate choice and sex-linked traits in other organisms have changed and evolved.

Box 2
Sexual Selection or Something Else?

We've discussed sexual selection several times in this book, first as an idea Charles Darwin used to explain the extreme or gaudy traits members of one sex (usually the male sex) of a species sometimes possess, for example, the peacock's long and bright tail feathers.[40] This trait may hinder the peacock's ability to avoid predators but will improve his chances of mating with peahens. Darwin also used sexual selection to explain other differences between the sexes, including why males tend to be larger than females in our own species. He thought human males competing for women needed stronger and bigger bodies to fight their competitors, saying:

> There can be little doubt that the greater size and strength of man, in comparison with woman, together with his broader shoulders, more developed muscles, rugged outline of body, his greater courage and pugnacity . . . these characters would, however, have been preserved or even augmented during the long ages of man's savagery, by the success of the strongest and boldest men, both in the general struggle for life and in their contests for wives; a success which would have ensured their leaving a more numerous progeny than their less favoured brethren.[41]

Darwin's explanation hasn't been called into question as often as you might expect; although, recently, a biological anthropologist, Holly Dunsworth, proposed an alternative explanation that doesn't require old gender stereotypes of savage men and small, meek women.[42] Dunsworth found that the differences we see in height and pelvic width in humans can be explained by simple hormonal differences between the sexes. In general, women will be shorter and have wider hips than men because human bone growth relies on the hormone estrogen, and people with ovaries produce more of this hormone during puberty (and less afterward) than those lacking ovaries do (which is why adolescent boys are generally shorter than girls but continue growing to a later age)—thus physiological

Box 2 (continued)

differences that have previously been almost exclusively attributed to sexual selection can be explained without a Victorian-age male-on-male wrestling competition story.

Although it can sometimes be overemphasized, sexual selection theory is extremely valuable for explaining the variation we see within and across species.[43] It rests upon the fact that, in most sexually reproducing animals, there is competition among the sexes for limited resources. We tend to describe sexual selection as just male competition, but it is more often better described from the perspective of "female choice." Think of the limited number of egg-producing individuals selecting their mates from among the more numerous sperm-producing resources (more numerous because sperm is more abundant, and males can produce sperm over a longer period of their life, while females have a shorter window for pregnancy); that's why males need flashier coloration or why they create more decorative nests to attract females (and, yes, to sometimes fight among themselves, too). On the other hand, there are also species where males are the limited resource and are the choosy ones, such as pipefishes, where males carry the developing eggs; or as in the combative female red phalarope (*Phalaropus fulicarius*), with her bright plumage in contrast to dull-colored males who dutifully incubate the eggs in her territory even after she flies off to begin her migration. Sexual selection is a powerful tool for explaining diversity across the Tree of Life; taking a view that isn't skewed by our biases may improve our understanding of the role of diversity in evolutionary biology.

COMBATTING POST-TRUTH WITH TRUST

We live in the "post-truth" era. People can say what they want, and some people, especially their followers, will believe it. I don't think you can combat "post-truth" with actual truth—people will continue to believe what they want, and we all have our own truths. What we need is trust. People believe those they trust. Maybe you believed in Santa Claus because you trusted your parents or in "trickle-down economics" because you trusted Ronald Reagan. I trust in science because I learned how to be a scientist from people who taught me to discover and explore on my own. And, from carrying it out, I learned to trust the scientific process.

I once had an undergraduate student in my Evolution class who didn't believe the moon landings actually happened. As I said in the beginning of this book (about other folks with similar views), I could have given her all the scientific papers that resulted from the lunar landings or shown her how we still use the reflectors left behind on the moon to measure distances with lasers, but I know she would have remained unconvinced. What I had to do was earn her trust, get her to trust science, and show her how to discover the truth on her own by finding and understanding the available research.

Something I think I'm able to do with many students but sadly not with her; she dropped out. I hope I have earned your trust, and convinced you to do what those other students of mine were able to do on their own, which is to read peer-reviewed scientific articles and convince themselves that what the "sage on the stage" professor was telling them was trustworthy.

What I teach my students are not just parades of facts that I expect them to regurgitate on an exam. I show them how to look up and understand scientific literature and to make up their own minds. And I have found that the facts my students find on their own are a hundred times more valuable and memorable than any I tell them.

At the same time, if you think about the things you believe are true, you probably didn't do a literature review or experiment to find confirmation. You already trust the sources and people from whom you learned these facts. If you think about it, you believe the moon landings happened, or that COVID-19 is deadly, because of who—and what—you trust. And the same is true for the subject of evolution. Whether or not I've gained your trust here, I do provide a few hundred references in this book to get you started examining that scientific literature. And as you've seen, the science also changes over time; Darwin didn't know about genetics and hormones and their role in our modern understanding of evolution and sex. These aren't changes in opinion or of interpretation; these changes are from the accumulation of new data over time from which old theories and hypotheses are reshaped or discarded.

Of course, scientists have opinions and biases, too. Our job is to keep our opinions and biases out of our work as best as we can and to let the data speak for itself. As I mentioned in the beginning of this book, teaching evolution helped me realize that this topic remains controversial because of trust or, more often, the lack of it. You can't have people trust your data without them trusting you and the scientific community at large. Some folks lost trust in scientists at the beginning of the COVID-19 pandemic because there was some waffling between whether or not wearing a cloth mask would be beneficial and because of a push to use hand sanitizer which in the end may not have been so useful against SARS-CoV-2. But those mistakes are also part of the trial-and-error of testing hypotheses that are also part of the scientific process. Trust the scientific process if nothing else: it has gotten us to where we are today.

Scientists aren't perfect, and pursuing scientific truth is sometimes easier than earning trust for us. Some of us act like we are the sole purveyors of truth and that all scientists should be trusted, although, as in all professions, there are some who should not be. Some scientists, by crossing the line from helpful critiques to biased, unfair criticism, have pushed others out of the field, often colleagues who are more junior or who are from underrepresented groups.[1]

We scientists need not only to improve communications between colleagues but also to better incorporate (not appropriate) the knowledge of indigenous people into our science: Andrea Reid and others call this incorporation of indigenous knowledge into mainstream science "Two-Eyed Seeing."[2] And incorporating local people into the scientific process can also increase people's trust in science and scientists, as can other interactive scientific outreach.[3] Perhaps a larger issue is that it often appears that just wealthy countries and

wealthy people do science. We need to build scientific infrastructure and reach out to communities where people can't,
or won't, come to universities or museums to show that science is not just for the privileged. We've done a poor job of
creating an inclusive and welcoming environment in science
and of fostering diversity in fields that study diversity. But
before we can spread knowledge, we need to *earn trust*.

How do you earn trust? I don't think there's any rule book
that works for everyone, but I do think one-on-one personal
communication and listening can help. (Something you
can't do over social media platforms.) Also *not* doing things
that would make someone *lose* trust in you because, once
lost, trust is almost impossible to get back.

I happen to like meeting new people, which is a good trait
to have for earning trust (and probably linked to why I'm
also trusting of others). I like to listen to the people I meet
as they explain to me what they do and how they got there.
I've met security guards, real estate agents, mothers, fathers,
billionaires, the unemployed, and all kinds of other folks.
Listening to people's stories and acknowledging their backgrounds, hardships, and passions can help you earn their
trust, especially if you are also willing to be open-minded
and open-hearted.

The folks I meet sometimes ask me what I do, and I say
some version of "I'm a dad of twins and a husband, and I
teach evolutionary biology and I'm a fish curator at Louisiana
State University." Then we talk about sports or something
along the lines of "What's a fish curator do?" Sometimes,
not always, I get: "Evolution—isn't that just a theory?" Some
folks are locked and loaded with that one. The rest of the
conversation usually goes something like the one I have in

the next chapter. I don't try to change people's minds during these conversations. If they are coming from a religious perspective, for some, nothing short of a collapse of their worldview would get them to accept evolution as explaining their origins. And that's true of me, too, only the other way around—my scientific worldview (the "background knowledge" from which I understand the world) would have to collapse for me *not* to accept evolution. When I'm trying to earn trust, mostly I try to explain as a teacher what evolution is—and what it *isn't*. I'm not trying to indoctrinate anyone; it doesn't matter if the particular people I'm with understand science or not (this isn't a classroom). I'd sure like them to understand, and I will try to find common ground, but I never look to start an argument. In these casual situations, I'm really trying to get them to open up about how they see the world, and, that way, maybe I'll learn something, too. Perhaps they might ask something I don't know the answer to, or give me a new perspective on their worldview. That's my real job as a scientist and science communicator: to gain and to share knowledge. I hope the conversation in the next chapter, which I wrote based on real questions and online comments I've gotten over the years, can be instructive. Who knows, maybe it can help you have a constructive conversation with someone who wants to learn more about evolution but is reluctant, or who is coming from a creationist's perspective.

CONVERSATION WITH A CREATIONIST

It would be nice if conversations would work like the one I've written below; it is unusual these days for people to get to finish their thought before being interrupted. I hope the real conversations you may have can be so constructive.

Meet average citizen "Joe," who is a creationist, or at least someone who "doesn't believe in evolution," and evolution professor "Pro"—who is me, but more articulate. They're sitting in a quiet hotel bar around closing time, watching a baseball game on TV. After a few minutes of small talk Joe asks what he's been itching to ask ever since learning that Pro teaches evolution.

JOE So, umm . . . isn't evolution just a theory?

PRO Well, frankly, no, it isn't. Evolution is a fact. Evolution is how we explain the origins and diversity of life. From what we have studied and can observe we know life on Earth has changed, in other words "evolved" (Pro puts up air quotes), one life form from another, all from a common ancestor about four billion years ago. The evidence includes everything from fossils, to the shared DNA of all of life, to the modified limbs and organs that make up our bodies.

JOE So why is it called the "theory" (Joe puts up air quotes) of evolution?

PRO There are lots of theories of evolution. The one that people are most familiar with is "natural selection": that's the one that Charles Darwin came up with. There have been other theories, some right, some wrong. But none of those ideas about how evolution works changes the fact that life on Earth evolved.

JOE So what's so special about Darwin's natural selection? Why is that a theory?

PRO Well, I think we're using the word "theory" differently. Scientists don't usually distinguish between "theory," "law," and "fact" in the same way we do in everyday conversation. Saying "theory" doesn't mean that the jury is still out. That's why I like using "fact" in everyday bar parlance. Not that I'm in a bar every day . . . sometimes I drink at home.

JOE Nice. Me too. But I don't get it, man, tell me where you think we came from.

PRO See, we are humans, that makes us apes, mammals, vertebrates, a small twig on the Tree of Life.

JOE The Tree of Life?

PRO Sorry, yes, the "Tree of Life" is the depiction of all life on Earth, living or extinct, showing how we organisms are all related to one another. So we are apes because we are in the part of the Tree with other ape species.

JOE I don't quite get it. Why are we apes? I don't think I came from a gorilla, although the hair on my back is pretty silvery these days.

PRO Fantastic, thanks for sharing that (*Pro sips from Manhattan, picks up pen, and draws a neat evolutionary family tree of primates, like the one in figure 3, on a bar napkin as he talks).* We didn't come from gorillas specifically, but we are apes. We share a common ancestor with chimpanzees from a few million years ago. And with gorillas a few million years before that. Chimps are our closest living relatives, but there were almost a dozen other "human" species that were more closely related to us that went extinct. If they were around, you could more easily see that transition between us and the rest of the apes. Where do you think we came from, Joe, if I may ask?

JOE I was made in God's image, just like the [insert favorite religious text] says. What religion are you?

PRO If it is okay with you, I'd like to keep my religious beliefs out of this conversation and keep my answers about our origins strictly scientific, at least on my side. Since you're asking me about the science of evolution and all.

JOE *(Joe sips beer, looks up at baseball game showing on the screen)* Don't you think all that you study is wrapped up in how we see things now, I mean we weren't here millions of years ago, so we have to make sense of old bones and make up stories. No one has seen those fossil humans walking around.

PRO That's what humans did for a long time, yes. Make up stories based on what they saw from the past.

JOE Maybe like God creating people in his image, and, in return, people creating God in their image.

PRO Sure . . . And that's an interesting perspective, Joe.

JOE So you're saying I'm right?

PRO I'm not here to tell you that you are right or wrong. But as human culture changed, we created more sophisticated tools to test those stories from the past. We can use DNA to see our place on the Tree of Life, and we have fossil bones of other humans that we can compare with the bones of other apes and animals. We even have the DNA from some of those fossilized extinct humans like Neanderthals and Denisovans.

JOE Denny-who's-van? You making this stuff up, man? How do you get DNA from fossils?

PRO That's a great question Joe. There are lots of human fossils, in part because apparently some of our close relatives liked to hide out, live, and be buried in caves.

JOE: You mean like the Taliban or something.

PRO Yes, exactly (*Pro laughs with smiling Joe*). Some of those extinct human species died out recently enough that their bones didn't turn to stone yet in those caves. They didn't fully decompose so we could get DNA out of them. That's why we can tell that people of European ancestry have around 2 percent Neanderthal DNA as part of their genome, and people of Asian ancestry have more Denisovan in them. Have you taken one of those DNA tests?

JOE Not yet, Doc, but I'm planning on it, except I know I'm 100 percent American, baby!

PRO That's right, man (*They clink glasses and break into a "USA! USA!" chant, startling the other bar patrons; Pro doesn't tell Joe he was born in Canada*).

PRO There are so many weird paradoxes. A lot of people think we humans have the most DNA, or the most genes, but it's much more complicated. If you stretched out the DNA in just one of any of our cells, it would be about as tall as we are; but if you took all the DNA from an amoeba, which only has one cell, the genome would be taller than this building. Rice and mice and some parasites have more functional genes than us. It's crazy.[1]

JOE Okay, so our DNA might not be all that complex compared to slugs or whatever. Tell me this, though: Can you name a species that has evolved from natural selection?

PRO How about *Homo sapiens*, our species. We had to struggle to survive and compete with other species, including other human species, that probably led to our bigger brains, advanced language skills, maybe even to innate beliefs.

JOE You think we beat out those other species because we believe in God?

PRO Not necessarily in a particular God, or gods, but maybe a shared belief system gave us an advantage over other human species that perhaps were solo or in smaller family groups. But now I'm fighting above my weight class a little bit. Let's just call that one of the theories of modern human evolution. Anyway, earlier you asked me about Darwin and I didn't answer you completely.

JOE Yeah, I was wondering about that.

PRO I actually just got back from the Galapagos Islands with an evolution class, and you can see what Darwin saw and think like he did and wonder, "Yeah, why is there a different kind of tortoise on each island?" Or like Kurt Vonnegut wondered, "Why is there a finch adapted to drink blood out here, instead of say, a vampire bat?"[2] And if God can put anything anywhere, why aren't there frogs on these islands? You can't come back from those islands and think, well, God put everything everywhere, instead you think, "Oh, there are no frogs here because they can't cross a salty ocean, and the finches and tortoises are different because they evolved and dispersed here and are adapted to the unique environment of each island."

JOE You teach all that stuff to college students?

PRO I show them these ideas, and I try to provide as much evidence as I can to let them make up their own minds. I'm not telling them to believe me—they can believe what they want. My job is to teach them what the available science is and to show them how to find and understand that evidence in the scientific literature.

JOE So how does someone find that scientific info?

PRO Glad you asked. You got your phone on you? (*Pro eats one of the cherries out of his nearly finished drink. Joe whips out his smartphone. Pro shows him how to look up papers on Google Scholar by typing in "human evolution," and "natural selection." He teaches Joe how to compare the methods and results in scientific papers with opinion pieces or fake news and commentaries. Pro buys Joe a beer and gives him a card to tell him to reach out if he has more questions or if he can't get access to a paper. Pro tells Joe to look up the journals* Evolution, Systematic Biology, *and* Science *for more about the topic. He then gives him a discount code for* Explaining Life through Evolution, *a new book he wrote that he hopes can help explain the basics of evolution to anyone who's interested.*)

EPILOGUE

This book is titled *Explaining Life through Evolution*, and I wanted "life" to mean "all of life on Earth" and maybe also your individual "life." So it seems fitting that I'm finishing up writing this in "The Age of Corona," when we are thinking about the fragility of our own lives in the greater scheme of things. On April 1, 2020, then–US President Donald Trump stated: "This is a very brilliant enemy. You know, it's a brilliant enemy. They develop drugs like the antibiotics. You see it. Antibiotics used to solve every problem. Now one of the biggest problems the world has is the germ [COVID-19] has gotten so brilliant that the antibiotic can't keep up with it."[1] Now Donald and I are just a couple of guys from Queens; but he's into real estate, and I'm a biologist, so I don't expect him to know as much biology as I do, although I do hope he will learn that antibiotics are for bacterial infections and that viruses are not bacteria. If you put that aside for a moment, his point is actually an evolutionary one. "The germ has gotten brilliant"—yes, the virus gets more and more "resistant" because of evolution, via natural selection. Even with a vaccine, we will still need to be cautious as the virus evolves.[2] The viruses that can't withstand the antivirals in a vaccine will die, but not all of the viruses die; the ones that survive make copies of themselves that will also survive those same antivirals in the next round—that's how they become "resistant."[3] Some survive and some don't because of the variation in the virus's genetic code.

At the time of this writing, COVID-19 has caused over one million deaths in the United States and has left nearly seven million dead worldwide. Those numbers are still growing and are an underestimate since there were many undiagnosed COVID patients who died early in the pandemic and weren't counted;[4] there are many millions who have suffered from long-term chronic illness caused by this coronavirus ("long COVID").[5] Some models predict that it benefits most viruses to be less virulent over time because if they kill their host, they stop the spread of their copies, but there are also viral diseases like rabies, which still has an almost 100 percent mortality rate in humans.[6] The problem is that this coronavirus is changing very rapidly and spreading very quickly and has a mortality rate reported to be as high as 4 percent, or four times higher than the flu).[7] Unlike the 1918 Spanish flu, which killed perhaps 100 million people, many of whom were twenty to forty years old,[8]* this new disease is targeting the old and generally sparing the young (although look out for new variants).

The fact is that it's still too early to know how the virus is going to evolve. (I hope those of you reading this in the

---

* If you want a dark history of one of the deadliest virus pandemics ever, the so-called Spanish flu (so-called because it definitely didn't start in Spain; the US Midwest is its more likely place of origin), read Laura Spinney's chilling and academic *Pale Rider*.

future, remember how scary it was in the thick of the pandemic.) We may get a new variant or, worse yet, an altogether new strain (kind of like a new virus species) that is more virulent. That's part of why those of us who understand evolution are wearing masks and protecting ourselves with social distancing, because we know how a coronavirus spreads (although many folks are also acting like "germ theory" is "just a theory").[9]* One of the remarkable things I've seen during the course of this pandemic is just how many people are thinking about evolution now because of the huge amount of data available to us. Many of us learned how this virus is spreading by looking at how it changes in its hosts (that's us) over time, and because it evolves so quickly, we can see it change in near real time on a phylogeny (a little disheartening to watch but nextstrain.org is tracing the tree of SARS-CoV-2 on a daily basis). I remember in the early days of the pandemic learning from the SARS-CoV-2 phylogeny that most of the initial infections in New York City actually came from multiple invasions from Europe and not from Asia, as many had assumed.[10]

Evolution has also taught us why we get so many new variants of this coronavirus: it's because viruses are very sloppy at copying themselves (RNA to our DNA back to RNA), and that makes for lots of mutations—enough mutations that the immune response that we have might not be

---

* Sonia Shah does an excellent job in her book *Pandemic* explaining the psychology behind people resisting germ theory. There is also a great section in Candice Millard's *Destiny of the Republic* that chronicles the changing behavior of medical practitioners at the end of the nineteenth century.

able to recognize a new version (you can be reinfected and get sick multiple times). That's why, despite the remarkable speed with which vaccines became available (thanks again to evolution research)[11] we have to play catch-up to make new vaccines that help our immune system recognize the new variants. SARS-CoV-2 changes and morphs (evolves), and if you don't have a prior history with it, you get very sick and may even die. In the face of that fact, some people have argued that we should *let* people die, whether from COVID-19 or from future pandemic diseases, to allow the "fittest" to survive (they are thinking of actual physical fitness, not reproductive success as evolutionary biologists use the term "fittest"); except in that case they are the ones who decide who is "fit" (clearly they do not mean the immune-compromised, frontline health workers, or the poor). So that's actually artificial selection, if not straight-up eugenics. Those are the folks throwing peanuts at the kid with the allergy or stomping on your glasses or coughing on nurses.

Trump would sometimes speak about what he called the "racehorse theory" at his political rallies, telling the nearly all-white crowd of his supporters at one rally, "You think we're so different? You have good genes in Minnesota." He also praised American industrialist and Hitler's hero, Henry Ford, for having "good bloodlines," both remarks being closely tied to eugenics and antisemitism.[12] But Trump is not alone in saying them: political leaders across the globe are also promoting eugenics-based pseudo-science ideas.[13]

Perhaps that's why the teaching of evolutionary biology is under serious threat (again) in the United States as some of the protections against teaching religion as an alternative to evolutionary biology in schools were weakened by recent

Supreme Court decisions (see the impacts of overturning parts of the Establishment Clause of the First Amendment, as well as the Lemon and Endorsement Tests).

How do you think people who favor eugenics and discredit evolution will use technology like CRISPR on human embryos? Or how will they use your genetic information in health screenings that determine your medical coverage or academic qualifications?

So, yes, understanding evolution matters—our lives and our future, may depend on it.

ACKNOWLEDGMENTS

Anne-Marie Bono at the MIT Press and Penguin Random House Editor Premanka Goswami helped make this book possible and gave me the kind of encouragement that I had always hoped to find in editors—I thank them profusely. Ethan Kocak beautifully illustrated my "Darwin movie," "From So Simple a Beginning," in the middle of this book; Ethan is awesome. I thank Louisiana State University, my employer, for the freedom to write what I think and for a job doing all the things I love anyway. The LSU Center for Collaborative Knowledge is an amazing group of humanities, arts, and science faculty, who truly gave me a different perspective on the concepts of nature and evolution through their Aristotle/Darwin Seminar reading group; many of our conversations made it into the pages of this book.

The people at TED gave me the opportunity and confidence to speak on evolution and the platform to be seen and heard; the TED Fellows team is amazing. My special thanks to Shoham Arad, Patrick D'Arcy, Tom Rielly, and others, past and present, on that team. And I want to thank Wikipedia and Google (especially Google Scholar), for their publicly available information, as well as the staff at the libraries

I've used—and to Alexandra Elbakyan for making many scientific publications available to those without means. My many thanks also to Mary Jane West-Eberhard, Doug Futuyma, Morgan Kelly, Jeremy Yoder, Deborah Goldgaber, and others for their helpful comments on various drafts— any remaining mistakes are obviously on me. And I suppose I should also thank SARS-CoV-2, if only for helping so many people better understand evolution, by forcing us to learn about infections, transmission, and mutation.

My 2020–2021 pandemic time was largely spent in Ottawa, where I worked remotely for LSU and as a Visiting Fulbright Chair at Carleton University; my thanks to them for the opportunity to work safely with my family. And thanks to Leslie Rissler of the National Science Foundation, who told me I should write a book on evolution, so I did: thanks, Leslie.

Most of all, I thank my family and especially my wife for spreading our genes, our children for carrying them, but also for their love and support. I can accomplish little without them. We had foster children with us for much of our time during this pandemic and while writing this book, and despite the chaos, they helped me think about the world very differently and with greater appreciation. (If you have privilege and resources, please consider fostering a child in need.) My parents, Chitta and Anurupa Chakrabarty, gave me life and taught me about life, so I could not write this book about life without their insights and chromosomes— thanks, Mom and Dad, I love you more than is required by evolution.

GLOSSARY

Allele—a version of a gene, often expressed as dominant or recessive in simplified form, such as in a dominant allele for purple flowers and a recessive allele for yellow flowers, where the dominant purple allele can mask the recessive yellow allele.

Allopatric speciation—the formation of new species (speciation) after a physical barrier splits a population of organisms in two; in this case, isolation leads to speciation in the geographically separated populations.

Anthropocene—what some prefer to call the most recent portion within our current geological ("Holocene") epoch; specifically, the period of time since humans began to destructively manipulate and reshape life on Earth. This term is used in recognition of human-caused extinctions, the replacement of wildlife with livestock, and the general transformation of the landscape, seascape and freshwaters by humans that, in the view of some, are akin to other massive biological or seismic geological shifts that defined epochs in the Earth's past.

Archaea—a domain of life (one of three, along with Bacteria and Eukaryota) that comprises mostly extremophile, superficially "bacteria-like" organisms, sometimes lumped in with bacteria as "prokaryotes" to differentiate them from the mostly multicellular eukaryotes.

Artificial selection—when humans co-opt natural selection by selectively breeding domesticated animals or by manipulating their genes for particular preferred traits. Humans have used artificial selection by breeding to get our various dog breeds, dairy cows that produce prodigious quantities of milk, and varieties of corn that grow much bigger and faster than their wild ancestors do. But artificial selection now also includes direct manipulation of the genotype of plants or animals to produce, for example, fruits with traits we prefer, like the nonbrowning Arctic Apple.[1]

Catastrophism—the idea that massive, sudden cataclysmic events, such as an asteroid striking the Earth or severe, continent-wide earthquakes, led to major changes in the history of the Earth. Once seen to be in opposition to Darwin's slow and gradual change ("uniformitarianism"), catastrophism was also the predominant geological worldview for "young-Earth" proponents before uniformitarianism explained how slow and gradual geological processes were the norm and how they pointed to Earth being a very, very old planet (see the works of James Hutton and Darwin's friend Charles Lyell[2]).

Coding region—also called the "reading frame," the part of a gene that is "read" to produce any given protein; and where a mutation can have disastrous consequences, as opposed to a "noncoding region" of DNA, which is sometimes referred to as "junk DNA," where a mutation may have only minor or neutral consequences. Our "junk DNA" includes some of the old broken no longer functional genes that are part of the legacy of our ancestors, which some call "fossil genes."

Creationist—someone who believes in a nonevolutionary, sometimes antiscience religious account of the origins of humans or of all life on Earth. Creationists come in many different varieties, including biblical literalists who believe that God first created a man, Adam, and told him to name all the animals on Earth (he was a taxonomist!), and then created a woman, Eve, from one of Adam's ribs.[3] Not all Christian creationists, however, are biblical literalists who believe in this Book of Genesis account of Creation, and there are many deeply religious people who have no problem fitting the narrative of their religion with a belief system that also accepts science.[4]

CRISPR—acronym for "clustered regularly interspaced short palindromic repeats," an immune mechanism with which some bacteria fight off infectious viruses by cutting out the introduced DNA of an attacking virus and replacing it with a benign sequence that helps them "remember" that virus in case it attacks again. CRISPR has been co-opted by scientists as a powerful genetic tool, known fully as "CRISPR-Cas" and sometimes called a "gene-editing typewriter" or "word processor," for changing or deleting the DNA sequences of genes.

Epigenetics—the study of how the environment influences and can cause a change in the phenotype of an organism without heritable changes to its DNA. Epigenetics explains why identical twins aren't exactly identical or why smokers may have some of their genes "shut off" because of damage to their DNA; but since these changes aren't coded into the DNA but rather involve how the DNA is read, they aren't passed down to the offspring of the organism.

Eugenics—the vile notion that humans can gain "favorable" traits through selective breeding (as in artificial selection); but who determines what is "favorable" and who is doing the selecting of potential partners makes all the difference. A couple picking healthy embryos for a planned pregnancy or someone selecting a mate with features that someone finds attractive is *not* eugenics. The term has been co-opted and "softened" by would-be eugenicists to make it more palatable to the general public. Don't be like them.

Eukaryotes—life forms of the domain Eukaryota, which comprises all multicellular and a handful of unicellular organisms, including plants, fungi, and animals—pretty much all living organisms but bacteria and archaea, whose cells typically have a single circular chromosome but no nucleus, whereas the cells of eukaryotes do have a nucleus, one containing multiple chromosomes.

Evolution—the process by which species are born and modified, leading to all known life. (For the purposes of this book, our definition refers to "organic evolution" (not the evolution of other things such as boy bands); but see also the more detailed definitions of "evolution" sprinkled throughout the book.

Gene drive—a genetic mechanism that can "fix" any gene variant (allele) by ensuring that it is spread to all individuals in a population of organisms. As co-opted by geneticists, a gene drive could make a trait like "HIV resistance" pass down 100 percent of the time to the offspring of a human who has this trait plus the gene-drive transmission mechanism. Gene drives could help us break the rules of Mendelian inheritance. But we haven't used them in humans yet because we don't know what would happen if their transmission mechanism jumped to another, nontarget species via a virus or some other vector; or to other nontarget traits, with possibly serious, even disastrous consequences (or by making us not just disease resistant but also tall and handsome). Gene drives have enormous potential, but using them wisely requires an abundance of caution.

Genetic drift—the process by which populations or species diverge because of nonadaptive changes in their DNA, including the accumulation of different neutral mutations. Genetic drift is perhaps the most significant contributor to genetic change and evolution.

Genome—the totality of DNA in an organism, usually referring to the DNA arranged in chromosomes in each of the organism's cells, while other genomes, like the "mitogenome," refer to all the DNA in the mitochondria of an organism.

Genotype—a descriptor referring to an organism's DNA as opposed to its "phenotype," which is the physical or behavioral representation of the organism based on its genotype, the genotype includes the organism's genes, genetic architecture, and genome.

Genotyping—the process by which someone's DNA is examined for a few thousand commonly variable markers of the human genome (far less than even 1 percent of the total genome is examined). Sometimes people say they had their "genome sequenced" when, in fact, they were only genotyped, usually by taking a commercial genealogy or DNA test. Your DNA can actually vary (because of mutations) anywhere in your genome compared to the genome of another human, but the markers examined in genotyping are used as a shortcut for helping us understand how your DNA generally compares to reference (fully sequenced) genomes.

Homologous—having heritable traits that are structurally similar and derived from the same common ancestor in two or more different species of organisms descended from that ancestor. Thus the fin of a dolphin, the wing of a bird, and the arm of a human—which all have the same pattern of these anterior limb bones—are all said to be "homologous," that is, they are all derived from the same common, in this case, vertebrate ancestor that first evolved those limb bones. On the other hand, the wing of a fly and the wing of a bird are considered "analogous," not "homologous," because, even though the wings of these two groups are in some ways similar, we can't trace their origins to the same common ancestor—thus fly and bird wings evolved independently.

Hypothesis—a proposed explanation of some natural phenomenon: for example, "Apples fall to the ground when dropped because of gravity"; or "Apples fall to the ground when dropped because wind pushes them to the ground." Note that both explanations are hypotheses; science is about testing to see which is the best explanation for the phenomenon. Scientists continually return to their hypotheses to reexamine and refine them based on rigorous testing and as new information and new technologies arise.

Life—all living things that we know of use a DNA code and would appear on the Tree of Life depicting the relationships of all those living things. NASA defines it like this: "Life is a self-sustaining chemical system capable of Darwinian evolution,"[5] but I would drop the "Darwinian" (which usually refers to "natural selection" specifically) and just say "capable of evolution." Except that individuals don't evolve, populations do. Also, what to do about viruses; are they "self-sustaining" if they require other types of organisms to replicate? I'm sticking with the definition I provided before (in Part III); life is defined as "a property possessed by any organism that can inherit or pass down heritable traits and that can potentially participate in evolution."

Macroevolution—large-scale evolutionary change over deep time; the sum of microevolutionary processes underlying, for example, the origin of whales or of flowering plants.

Meiosis—the formation of sex cells (eggs and sperm); in humans, who typically have two sets or pairs of chromosomes; the sperm or egg gets a unique mix represented in a single chromosome set. During meiosis, a single cell divides twice to make four daughter cells (each with half the chromosome number as the parent cell). The daughter cells (egg and sperm) that combine during reproduction again have two sets of chromosomes formed from that union.

Microevolution—small-scale evolution such as genetic drift or natural selection acting on populations of organisms, typically taking place over relatively short periods of time, on a generational rather than a geological time scale.

Mutations—at the level of the genotype, a mutation is any change in an organism's DNA sequence that the organism did not inherit.

Natural group—a "monophyletic" group of organisms that includes the ancestor and all its descendants, such as "birds" when it is represented by *Archaeopteryx* and all other birds living and extinct descended from their common ancestor. However, because birds are a lineage of dinosaurs, you couldn't talk about dinosaurs being a natural/monophyletic group unless you also included birds. Monophyletic groups are real units of evolution because of their shared history, as opposed to, say, "bugs" in the colloquial sense, which may include a wide variety of species, from "true bugs" (or Hemiptera, which are insects) to "pill bugs" (which are actually crustaceans) and other small invertebrate critters that are only distantly related.

Natural selection—the process by which individual organisms compete or struggle for survival in a given environment; those which are fittest will survive and reproduce to give rise to the next generation. Natural selection was "codiscovered" by Charles Darwin and Alfred Russel Wallace, but it was made famous and described more thoroughly by Darwin in his *On the Origin of Species*.

Neutral evolution—also called the "neutral theory of molecular evolution" and championed by Motoo Kimura, the hypothesis that most genetic mutations do nothing to the phenotype; therefore, most evolutionary change is neutral.

Phenotype—the outward expression of the genotype of an organism, including everything from the proteins coded for by the organism's DNA to its body and behavior.

Phylogenetics—the study of relationships on the Tree of Life, which is an evolutionary depiction of the relationships of all living things on Earth.

Point mutation—a single mutation (change) in the DNA sequence of an organism, including a change from one DNA base to another, for instance, from an A to a C (adenine to cytosine) or a deletion or insertion of any one DNA base (A, C, T, or G—adenine, cytosine, thymine, or guanine).

Polymorphism—a natural state of variation for a given trait in a population of organisms, for instance, blond, black, brown, or red hair.

Polyploidy—the formation of more than the typical number of chromosomes for an individual organism or species. Having a genome duplication event can result in polyploidy, which can cause speciation or evolutionary innovations. After a genome duplication, an organism may end up with duplicate sets of genes that do identical tasks, so that one set can be freed up to evolve to do something else.

Recombination—the act of mixing the DNA from separate sets of chromosomes for daughter cells, which will each get half the original number of chromosomes. In the formation of sex cells (sperm and eggs) during meiosis, a parent with two sets of chromosomes produces these sex cells by mixing and matching—recombining—the two chromosome sets into a single set. Recombination is a major source of both variation and loss of variation in sexually reproducing organisms.

Sexual selection—a theory introduced by Charles Darwin in *On the Origin of Species* that explains why members of one sex of some sexually reproducing organisms (usually the males in vertebrates) have conspicuous traits (like bright colors or a loud call) that might threaten their survival by making them more susceptible to predation but that also make them more attractive to potential mates.

Sister group—a pair of closest relatives on the Tree of Life; "sister groups" are how the relationships of organisms on the bifurcating tree are depicted. Why not just say "closest relatives"? Because the true closest relatives of either of the descendants would be that shared common ancestor, which we generally do not know; these common ancestors remain unknown and are represented by "nodes" on the tree, which remain unnamed. Therefore, by saying "sister groups," we are acknowledging that uncertainty and recognizing the limits to how we can reconstruct the past.

Speciation—the evolutionary processes by which new species arise from an ancestor.

Species—population or populations of interbreeding or potentially interbreeding organisms, or a distinct evolutionary lineage with some heritable features that make its organisms distinct from those of other such lineages. What a species is exactly remains a hotly debated topic. And even though the many different concepts of species ("evolutionary species concept," "phylogenetic species concept," and "biological species concept," among dozens more) all agree that species are the products of evolution, there is no one-size-fits-all definition of species that works across the diversity of life forms.

Sympatric speciation—the formation of new species by factors other than the presence of physical or geographic barriers; isolation in sympatric speciation can be caused by factors such as premating assortment. where, for instance, females may be attracted to particular types of males before mating, or postmating factors, where certain males' genes do not contribute to the next generation; mechanisms like polyploidy may also explain sympatric speciation in the absence of physical barriers.

Taxonomist—someone who describes and names new species or lineages and who also organizes the biodiversity of life forms within a hierarchy of names.

Theory—an overarching explanation for some natural phenomenon, not "a tenuous idea," as "theory" is used in everyday, nonscientific speech, but more like "law," as used in scientific language. As they do with their hypotheses, however, scientists continually return to their theories to reexamine and refine them on the basis of rigorous testing.

Tree of Life—the depiction of the relationships between living things on Earth as a series of (usually bifurcating) ancestor-descendant relationships. Put more simply, the Tree of Life is a diagram showing which species are related to which other species.

Uniformitarianism—the idea that slow and gradual change in organisms or geological formations can lead to large differences over time, such as continental drift across oceans. Although the continents move at the speed your fingernails grow, their movements over a 100 million years can add up to thousands of miles. Similarly, a population of tortoises separated into two populations by the formation of a barrier, such as a lava flow, or by some of its members dispersing onto a new island can accumulate enough changes in one or both of the populations over thousands of years to become different species. (See also *catastrophism*.)

Viruses—seen by most scientists as peripheral to the three domains of living things they all agree upon—Eurkaryota (or Eukarya), Bacteria, and Archaea—because viruses need other organisms to make copies of (replicate) themselves. Are they living things—part of life on Earth—or are they not? Depends on how you define "life." Think of viruses as being like a parasitic vine growing on the Tree of Life and drawing its sustenance from the branches of the tree as it wraps itself around them, but it can't live on its own without its host tree—and neither can viruses without their hosts. So perhaps they are sometimes alive, and other times not; but they are evolving biological entities nevertheless, and part of the evolutionary history of life on Earth.[6]

Vouchers—in the context of biological research, vouchers are preserved reference specimens of species usually stored in natural history museum collections. They can be used to verify the identification of species or as a reference in various kinds of biological studies, from providing evidence of shifting levels of pollution to documenting extinctions.

NOTES

CITATION DIVERSITY STATEMENT

Perry Zurn, Danielle S. Bassett, and Nicole C. Rust suggest that authors include "Citation Diversity Statements" such as this one to combat the negative biases that result in under citing women and people from other groups underrepresented in the sciences.* In an effort to make science more equitable, I've tried to include references that reflect the diversity of evolutionary biologists and others in the life sciences. My apologies in advance if, despite my best efforts; I've overlooked some important references.

PART I

CALLED TO ACTION

1. L. S. Mead and A. Mates, "Why Science Standards Are Important to a Strong Science Curriculum and How States Measure Up," *Evolution: Education and Outreach* 2 (2009): 359–371, https://evolution-outreach.biomedcentral.com/articles/10.1007/s12052-009-0155-y.

2. J. Rayfield, "Bobby Jindal: I'm Fine with Teaching Creationism in Public Schools," *Salon*, April 16, 2013, https://www.salon.com/2013/04/16/bobby_jindal_im_fine_with_teaching_creationism_in_public_schools/.

---

* P. Zurn, D. S. Bassett, and N. C. Rust, "The Citation Diversity Statement: A Practice of Transparency, a Way of Life," *Trends in Cognitive Sciences* 24, no. 9 (2020): 669–672.

3. Mi-yeon Eom, "The Lotus: Rebirth of the Self," *Journal of Symbols & Sandplay Therapy* 3, no. 1 (2012): 95–107.

4. S. J. Gould, *Rocks of Ages: Science and Religion in the Fullness of Life* (New York: Ballantine Books, 2002).

5. M. E. Barnes and S. E. Brownell, "A Call to Use Cultural Competence When Teaching Evolution to Religious College Students: Introducing Religious Cultural Competence in Evolution Education (ReCCEE)," *CBE: Life Sciences Education* 16, essay 4 (2017).

6. G. Branch, "Alabama Retains Its Evolution Disclaimer," *National Center for Science Education*, March 15, 2016, https://ncse.ngo/news /2016/03/alabama-retains-its-evolution-disclaimer-0016968https:// ncse.ngo/news/2016/03/alabama-retains-its-evolution-disclaimer -0016968.

7. J. Timmer, "Arizona State Education Standards See Evolution Deleted: Never Mention 'Change over Time' by Its Name," *ARS Technica*, May 23, 2018, https://arstechnica.com/science/2018/05/arizon ,a-official-waters-down-states-science-education-standards/.

8. E. Plutzer, G. Branch, and A. Reid, "Teaching Evolution in U.S. Public Schools: A Continuing Challenge," *Evolution: Education and Outreach* 13, article 14 (2020), https://doi.org/10.1186/s12052-020-00126-8.

9. J. Bell, L. Lugo, A. Cooperman, N. Sahgal, J. H. Martinez, B. Mohamed, et al., *The World's Muslims: Religion, Politics and Society* (Washington, DC: Pew Research Center, April 13, 2013), chap. 7: "Religion, Science and Popular Culture," http://www.pewforum.org /2013/04/30/the-worlds-muslims-religion-politics-society-science-and -popular-culture/.

10. P. Kingsley, "Turkey Drops Evolution from Curriculum, Angering Secularists," *New York Times*, June 23, 2017.

11. M. Le Page, "Evolution Myths: It Doesn't Matter If People Don't Grasp Evolution, *New Scientist* 198, no. 2652 (April 19, 2008): 31, https://doi.org/10.1016/S0262-4079(08)60984-7.

12. A. Jogelekar, "Monday Morning Levity: Louisiana Senator Asks If *E. coli* Evolve into Persons, Scientific American (blog), posted January 21, 2013, https://blogs.scientificamerican.com/the-curious-wavefunc tion/monday-morning-levity-louisiana-senator-asks-if-e-coli-evolve -into-persons/.

13. J. B. Losos, *Improbable Destinies: Fate, Chance, and the Future of Evolution* (New York: Penguin, 2017).

14. Z. D. Blount, R. E. Lenski, and J. B. Losos, "Contingency and Determinism in Evolution: Replaying Life's Tape," *Science* 362, no. 655 (2018: review summary) and *Science* 362 (2018): eaam5979 (full review article); R. E. Lenski, "The *E. coli* Long-Term Experimental Evolution Project Site" (2020), https://lenski.mmg.msu.edu/ecoli/index .html; Losos, *Improbable Destinies*.

PULLED INTO THE FIGHT

1. B. Forrest and P. R. Gross, *Creationism's Trojan Horse: The Wedge of Intelligent Design* (New York: Oxford University Press, 2004), https:// doi.org/10.1093/acprof:oso/9780195157420.001.0001.

2. A. Heist, "The Teenage 'Troublemaker' Fighting for Science," *Science Friday*, NPR, April 12, 2013, https://www.sciencefriday.com/seg ments/the-teenage-troublemaker-fighting-for-science/.

3. D. J. Futuyma, *Evolution* (Sunderland, MA: Sinauer Associates, 2005).

4. H. Zeberg and S. Pääbo, "The Major Genetic Risk Factor for Severe COVID-19 Is Inherited from Neanderthals," *Nature* 587 (2020): 610–612, https://doi.org/10.1038/s41586-020-2818-3.

5. S. Farina and M. Gibbons, "'The Last Refuge of Scoundrels': New Evidence of E. O. Wilson's Intimacy with Scientific Racism," *Science for the People*, February 1, 2022, ttps://magazine.scienceforthepeople.org /online/the-last-refuge-of-scoundrels/; M. R. McLemore, "The Complicated Legacy of E. O. Wilson," *Scientific American*, posted online December 29, 2021, https://www.scientificamerican.com/article/the -complicated-legacy-of-e-o-wilson/; M. Borrello and D. Sepkoski, "Ideology as Biology," *New York Review*, February 5, 2022; C. B. Ogbunu, "Ghosts of Science Past Still Haunt Us. We Can Put Them to Rest." *Undark*, posted online December 13, 2022, https://race.undark.org /articles/ghosts-of-science-past-still-haunt-us-we-can-put-them-to-rest.

6. E. C. Scott, *Evolution vs. Creationism: An Introduction*, 2nd ed. (Santa Barbara, CA: ABC-CLIO, 2008).

ON TRUST

1. Barnes and Brownell, "A Call to Use Cultural Competence"; L. J. Rissler, S. I. Duncan, and N. M. Caruso, "The Relative Importance of

Religion and Education on University Students' Views of Evolution in the Deep South and State Science Standards across the United States," *Evolution: Education and Outreach* 7, article 24 (2014), https://doi.org /10.1186/s12052-014-0024-1.

FACTS AND TRUTH

1. H. Allcott and M. Gentzkow, "Social Media and Fake News in the 2016 Election," *Journal of Economic Perspectives* 31, no. 2 (2017): 211–236.

2. J. Nielsen, R. B. Hedeholm, J. Heinemeier, P. G. Bushnell, J. S. Christiansen, J. Olsen, et al., "Eye Lens Radiocarbon Reveals Centuries of Longevity in the Greenland Shark (*Somniosus microcephalus*)," *Science* 353, no. 6300 (2016): 702–704.

3. M. J. Van Kranendonk, P. Philippot, K. Lepot, S. Bodorkos, and R. Pirajno, "Geological Setting of Earth's Oldest Fossils in the c. 3.5 Ga Dresser Formation, Pilbara Craton, Western Australia," *Precambrian Research* 167 (2008): 93–124.

4. A. Einstein, "Ist die Trägheit eines Körpers von seinem Energie-inhalt abhängig?" *Annalen der Physik* 18, article 13 (1905): 639–643, https://doi.org/10.1002/andp.19053231314.

5. S. Wolfson, "'Germs Are Not a Real Thing': Fox News Host Says He Hasn't Washed Hands in 10 Years," *Guardian*, February 11, 2019, https://www.theguardian.com/media/2019/feb/11/germs-are-not -real-fox-news-host-pete-hegseth.

6. C. Gohd, "'Mad' Flat-Earther to Launch Himself 5,000 Feet Up on a Homemade Steam Rocket Sunday," Space.com, August 9, 2019, https://www.space.com/flat-earther-mad-mike-hughes-august-2019 -launch.html.

PART II

INTRODUCTION TO EVOLUTION

1. J. Clack, "The Emergence of Early Tetrapods," *Palaeogeography, Palaeoclimatology, Palaeoecology* 232, nos. 2–4 (2006): 167–189, https:// www.sciencedirect.com/science/article/abs/pii/S003101820500444X ?via%3Dihub.

2. R. L. Carroll, *Vertebrate Paleontology and Evolution* (New York: W. H. Freeman, 1988).

3. M. M. Gilbert, E. Snively, and J. Cotton, "The Tarsometatarsus of the Ostrich *Struthio camelus*: Anatomy, Bone Densities, and Structural Mechanics," *PLOS ONE* 11, no. 3 (2016): e0149708; T. X. Qiu, E. C. Teo, Y. B. Yan, and W. Lei, "Finite Element Modeling of a 3D Coupled Foot-Boot Model," *Medical Engineering & Physics* 33, no. 10 (2011): 1228–1233.

4. S. J. Gould, *The Panda's Thumb: More Reflections in Natural History* (New York: W. W. Norton: 1980).

5. N. Shubin, *Your Inner Fish* (New York: Pantheon, 2008).

6. R. Diogo and J. M. Ziermann, "Muscles of Chondrichthyan Paired Appendages: Comparison with Osteichthyans, Deconstruction of the Fore-Hindlimb Serial Homology Dogma, and New Insights on the Evolution of the Vertebrate Neck," *Anatomical Record* 298, no. 3 (2014): 513–530, https://doi.org/10.1002/ar.23047.

7. A. Pradel, J. Maisey, P. Tafforeau, R. H. Mapes, and J. Mallatt, "A Palaeozoic Shark with Osteichthyan-Like Branchial Arches," *Nature* 509 (2014): 608–611, https://doi.org/10.1038/nature13195; M. D. Brazeau, S. Giles, R. P. Dearden, A. Jerve, Y. Ariunchimeg, E. Zorig, et al., "Endochondral Bone in an Early Devonian "Placoderm" from Mongolia," *Nature Ecology & Evolution* 4 (2020): 1477–1424, https://doi.org/10.1038/s41559-020-01290-2.

8. D. Bernal, C. Sepulveda, O. Mathieu-Costello, and J. B. Graham, "Comparative Studies of High Performance Swimming in Sharks, 1. Red Muscle Morphometrics, Vascularization and Ultrastructure," *Journal of Experimental Biology* 206, no. 16 (2008): 2831–2843.

9. F.-C. Chen and W.-H. Li, "Genomic Divergences between Humans and Other Hominoids and the Effective Population Size of the Common Ancestor of Humans and Chimpanzees," *American Journal of Human Genetics* 68, no. 2 (2001): 444–456, https://doi.org/doi:10.1086/318206.

10. V. R. Paixão-Côrtes, L. H. Viscardi, F. M. Salzano, R. Hünemeier, and M. C. Bortolini, "*Homo sapiens, Homo neanderthalensis* and the Denisova Specimen: New Insights on Their Evolutionary Histories Using Whole-Genome Comparisons," *Genetics and Molecular Biology* 35, no. 4

suppl. (2012): 904–911, https://doi.org/doi:10.1590/s1415-47572012 000600003.

11. W. E. Banks, F. d'Errico, A. T. Peterson, M. Kageyama, A. Sima, and M. F. Sánchez-Goñi, "Neanderthal Extinction by Competitive Exclusion," *PLOS ONE* 3, no. 12 (2008): e3972.

12. L. Hug, B. Baker, K. Anantharaman, C. T. Brown, A. J. Probst, C. J. Castelle, et al., "A New View of the Tree of Life," *Nature Microbiology* 1, article 16048 (2016), https://doi.org/10.1038/nmicrobiMol.2016.48.

13. D. Quammen, *The Reluctant Mr. Darwin: An Intimate Portrait of Charles Darwin and the Making of His Theory of Evolution*, Great Discoveries (New York: W. W. Norton, 2007).

14. J. François, "Evolution and Tinkering," *Science* 196 no. 4295 (1977), 1161–1166.

THE GREAT CHAIN OF BEING OR LADDER OF LIFE

1. M. Ragan, "Trees and Networks before and after Darwin," *Biology Direct* 4, article 43 (2009); T. W. Pietsch, *Trees of Life: A Visual History of Evolution* (Baltimore: Johns Hopkins University Press, 2012), 376.

2. J. B. Peterson, *12 Rules for Life: An Antidote to Chaos* (New York: Penguin Random House, 2018).

3. D. W. Bruce, "Serotonin in Pineapple," *Nature* 188, no. 4745 (1960): 147; P. Z. Myers, "PZ Replies to the Lobsterians," Youtube.com, February 11, 2018, https://www.youtube.com/watch?v=Sqx57l781WM.

4. B. Steinworth, "Jordan Peterson Needs to Reconsider the Lobster," *Washington Post*, June 4, 2018.

5. M. Ramenofsky, "Behavioral Endocrinology of Migration," in *Encyclopedia of Animal Behavior*, ed. M. D. Bread and J. Moore (Cambridge, MA: Academic Press, 2010), 191–199, https://doi.org/10.1016/B978-0-08-045337-8.00252-7.

6. Peterson, *12 Rules for Life*.

7. J. D. Monk, E. Giglio, A. Kamath, M. R. Lambert, and C. E. McDonough, "An Alternative Hypothesis for the Evolution of Same-Sex Sexual Behaviour in Animals," *Nature Ecology & Evolution* 3, no. 12 (2019): 1622–1631; J. A. Long, *Dawn of the Deed: The Prehistoric Origins of Sex* (Chicago: University of Chicago Press, 2012).

8. D. Aubert-Marson, "Sir Francis Galton: Le fondateur de l'eugénisme," *Médecine/Science* (Paris) 25, nos. 6–7 (2009): 641–645, https://doi.org/10.1051/medsci/2009256-7641.

THEORIES OF EVOLUTION

1. C. Darwin, *On the Origin of Species by Means of Natural Selection, or Preservation of Favoured Races in the Struggle for Life* (London: John Murray, 1859).

2. E. Mayr, *What Evolution Is* (New York: Basic Books, 2001).

3. C. Darwin, *The Descent of Man and Selection in Relation to Sex* (New York: D. Appleton, 1896).

4. M. Kimura, *The Neutral Theory of Molecular Evolution* (Cambridge: Cambridge University Press, 1983).

5. D. B. Paul, "The Selection of the 'Survival of the Fittest,'" *Journal of the History of Biology* 21 (1988): 411–424, https://doi.org/10.1007/BF00144089.

6. D. B. Paul, "Darwin, Social Darwinism and Eugenics," in *The Cambridge Companion to Darwin*, ed. J. Hodge and G. Radick (Cambridge: Cambridge University Press, 2003), 214.

7. R. M. Dennis, "Social Darwinism, Scientific Racism, and the Metaphysics of Race," *Journal of Negro Education* 64, no. 3 (1995): 243–252.

8. R. O. Prum, *The Evolution of Beauty: How Darwin's Forgotten Theory of Mate Choice Shapes the Animal World—and Us* (New York: Anchor Books/Doubleday, 2017).

9. Darwin, *On the Origin of Species* (1859 edition).

FROM SO SIMPLE A BEGINNING

1. C. Darwin, *The Autobiography of Charles Darwin, 1809–1882: With Original Omissions Restored* (New York: W. W. Norton, 1993); Quammen, *The Reluctant Mr. Darwin*; E. J. Browne, *Charles Darwin: Voyaging: Volume 1 of a Biography* (New York: Random House, 2010); E. J. Browne, *Charles Darwin: The Power of Place*, vol. 2 (Princeton: Princeton University Press, 2003).

2. R. B. Freeman, "Darwin's Negro Bird-Stuffer," *Notes and Records of the Royal Society of London* 33, no. 1 (1978): 83–86.

THE SELECTION OF NATURAL SELECTION

1. Darwin, *On the Origin of Species* (1859 edition).

2. E. Lurie, "Louis Agassiz and the Races of Man," *Isis* 45, no. 3 (1954): 227–242.

3. J. B. A. de M. Lamarck, *Philosophie zoologique, ou, Exposition des considérations relatives à l'histoire naturelle des animaux* (Paris: F. Savy, 1873).

4. A. Weismann, *Essays upon Heredity and Kindred Biological Problems* (Oxford: Clarendon Press, 1891), 1.

5. E. J. Sobo, "What Is Herd Immunity, and How Does It Relate to Pediatric Vaccination Uptake? U.S. Parent Perspectives," *Social Science & Medicine* 165 (2016): 187–195, https://doi.org/10.1016/j.socscimed.2016.06.015.

6. O. J. Watson, G. Barnsley, J. Toor, A. B. Hogan, P. Winskill, and A. C. Ghani, "Global Impact of the First Year of COVID-19 Vaccination: A Mathematical Modelling Study," *Lancet Infectious Diseases* 22, no. 9 (2022): 1293–1302.

7. R. Chambers, *Vestiges of the Natural History of Creation* (London: John Churchil, 1844).

8. A. R. Wallace, *The Malay Archipelago* (New York: Harper, 1869).

9. P. Raby, *Alfred Russel Wallace: A Life* (Princeton: Princeton University Press, 2002).

10. Raby, *Alfred Russel Wallace*.

11. C. Darwin and A. Wallace, "On the Tendency of Species to Form Varieties; and on the Perpetuation of Varieties and Species by Natural Means of Selection," *Zoological Journal of the Linnean Society* 3, no. 9 (1858): 46–62.

12. Darwin, *On the Origin of Species* (1859 edition).

13. Darwin and Wallace, "On the Tendency of Species"; J. Gleick, *Isaac Newton* (New York: Random House, 2004); S. R. Starr, *Lost Enlightenment: Central Asia's Golden Age from the Arab Conquest to Tamerlane* (Princeton: Princeton University Press, 2013); A. H. Malik, J. M. Ziermann, and R. Diogo, "An Untold Story in Biology: The Historical Continuity of Evolutionary Ideas of Muslim Scholars from the 8th Century to Darwin's Time," *Journal of Biological Education* 52, no.

1 (2018): 3–17; R. Stott, *Darwin's Ghosts: The Secret History of Evolution* (London: Bloomsbury, 2012).

14. C. Zimmer, S*he Has Her Mother's Laugh: The Powers, Perversions, and Potential of Heredity* (New York: Penguin, 2018); M. J. West-Eberhard, *Developmental Plasticity and Evolution* (New York: Oxford University Press, 2003).

MENDEL AND THE MAINTENANCE OF VARIATION

1. D. Galton, "Did Darwin Read Mendel?" *QJM: An International Journal of Medicine* 102, no. 8 (2009): 587–589, https://doi.org/10.1093/qjmed/hcp024.

2. P. Vorzimmer, "Charles Darwin and Blending Inheritance," *Isis* 54, no. 3 (1963): 371–390, https://doi.org/10.1086/349734.

3. U. Deichmann, "Gemmules and Elements: On Darwin's and Mendel's Concepts and Methods in Heredity," in *Darwinism, Philosophy, and Experimental Biology*, ed. U. Deichmann and A. S. Travis (Berlin, Springer, 2010), 31–58.

4. Deichmann, "Gemmules and Elements"; K. Kampourakis, "Mendel and the Path to Genetics: Portraying Science as a Social Process," *Science & Education* 22 (2013): 293–324, https://doi.org/10.1007/s11191-010-9323-2; R. A. Fisher, "Has Mendel's Work Been Rediscovered?' *Annals of Science* 1, no. 2 (1936): 115–137, https://doi.org/10.1080/00033793600200111; N. C. Strenseth, L. Andersson, and H. E. Hoekstra, "Gregor Johann Mendel and the Development of Modern Evolutionary Biology," *Proceedings of the National Academy of Sciences* 119, no. 30 (2022): e2201327119; L. Urry, M. Cain, S. Wasserman, P. Minorsky, and P. Reece, *Campbell Biology*, 11th ed. (London: Pearson, 2017).

5. J. F. Crow, "Advantages of Sexual Reproduction," *Developmental Genetics* 15, no. 3 (1994): 205–213, https://doi.org/10.1002/dvg.1020150303.

6. Galton, "Did Darwin Read Mendel?"; R. Olby and P. Gautrey, "Eleven References to Mendel before 1900," *Annals of Science* 24, no. 1 (1968): 7–20, https://doi.org/10.1080/00033796800200021.

7. J. Huxley, *Evolution: The Modern Synthesis* (New York: Harper & Brothers, 1943).

8. Huxley, *Evolution*.

9. Urry et al., *Campbell Biology*, 11th ed.; M. J. Ferreiro, C. Pérez, M. Marchesano, S. Ruiz, A. Caputi, P. Aguilera, et al., "*Drosophila melanogaster* White Mutant w[1118] Undergo Retinal Degeneration," *Frontiers in Neuroscience* 11, article 732 (2018); O. Vecerek, "Johann Gregor Mendel as a Beekeeper," *Bee World* 46, no. 3 (1965): 86–96, https://doi.org /10.1080/0005772x.1965.11095345.

MUTANTS AND MUTATIONS

1. J. C. Venter, M. D. Adams, E. W. Myers, F. W. Li, R. J. Mural, G. G. Sutton, et al., "The Sequence of the Human Genome," *Science* 291, no. 5507 (2001): 1304–1351, https://science.sciencemag.org/content /291/5507/1304.

2. M. Kimura, "Evolutionary Rate at the Molecular Level," *Nature* 217, no. 5129 (1968): 624–626.

3. Kimura, "Evolutionary Rate."

4. Kimura, *Neutral Theory*.

5. J. Gusella, N. Wexler, P. Conneally, S. L. Naylor, M. A. Anderson, R. E. Tanzi, et al., "A Polymorphic DNA Marker Genetically Linked to Huntington's Disease," *Nature* 306 (1983): 234–238, https://doi.org /10.1038/306234a0.

6. W. Stephen, W., "Genetic Hitchhiking versus Background Selection: The Controversy and Its Implications," *Philosophical Transactions of the Royal Society, B: Biological Sciences* 365, no. 1544 (2010), 1245–1253, http://doi.org/10.1098/rstb.2009.0278.

7. R. Shiang, L. M. Thompson, Y. Z. Zhu, D. M. Church, T. J. Fielder, M. Bocian, et al., "Mutations in the Transmembrane Domain of FGFR3 Cause the Most Common Genetic Form of Dwarfism, Achondroplasia," *Cell* 78, no. 2 (1994): 335–342, https://doi.org/10.1016 /0092-8674(94)90302-6.

8. M. J. Lindhurst, J. C. Sapp, J. K. Teer, J. J. Johnston, E. M. Finn, K. Peters, et al., "A Mosaic Activating Mutation in AKT1 Associated with the Proteus Syndrome," *New England Journal of Medicine* 365, no. 7 (2011): 611–619, https://doi.org/10.1056/NEJMoa1104017.

9. E. V. Ignatiev, V. G. Levitsky, N. S. Yudin, M. P. Moshkin, and N. A. Kolchanov, "Genetic Basis of Olfactory Cognition: Extremely High

Level of DNA Sequence Polymorphism in Promoter Regions of the Human Olfactory Receptor Genes Revealed Using the 1000 Genomes Project Dataset," *Frontiers in Psychology* 5 (2014): 247, https://doi.org /10.3389/fpsyg.2014.00247.

10. N. Eriksson, S. Wu, C. B. Do, A.K. Kiefer, J. Y. Tung, J. L. Mountain, et al., "A Genetic Variant Near Olfactory Receptor Genes Influences Cilantro Preference," *Flavour* 1, no. 22 (2012), https://doi.org/10.1186 /2044-7248-1-22.

11. B. McClintock, "The Origin and Behavior of Mutable Loci in Maize," *Proceedings of the National Academy of Sciences* 36, no. 6 (1950): 344–355, https://doi.org/10.1073/pnas.36.6.344; L. A. Pray, "Transposons: The Jumping Genes," *Nature Education* 1, no. 1 (2008): 204.

12. L. Sagan, "On the Origin of Mitosing Cells," *Journal of Theoretical Biology* 14, no. 3 (1967): 225–274.

13. D. Quammen, *The Tangled Tree: A Radical New History of Life* (New York: Simon & Schuster, 2018).

14. G. Cornelis, M. Funk, C. Vernochet, F. Leal, O. A. Tarazona, G. Meurice, et al., "An Endogenous Retroviral Envelope Syncytin and Its Cognate Receptor Identified in the Viviparous Placental *Mabuya* Lizard," *Proceedings of the National Academy of Sciences* 114, no. 51 (2017): 10991–11000, https://doi.org/10.1073/pnas.1714590114; V. Racaniello, "Retroviruses Turned Egg-Layers into Live-Bearers," Virology blog: About Viruses and Viral Diseases, posted December 14, 2017, http://www.virology.ws/2017/12/14/a-retrovirus-gene-drove-emer gence-of-the-placenta/; F. P. Ryan, "Viral Symbiosis and the Holobiontic Nature of the Human Genome, *APMIS* 124, nos. 1–2 (2016): 11–19, https://doi.org/10.1111/apm.12488; ""How Virus Infections Shaped Human Evolution," Youtube.com, March 27, 2019, https:// www.youtube.com/watch?v=nWuV6PVKv1A. Many thanks to Jeet Sukumaran for spreading the word of this cool story to me.

15. G. Petrosino, G. Ponte, M. Volpe, I. Zarrella, F. Ansaloni, C. Langella, et al., "Identification of LINE Retrotransposons and Long Non-Coding RNAs Expressed in the Octopus Brain," *BMC Biology* 20, article 116 (2022).

16. F. P. Ryan, "Viral Symbiosis and the Holobiontic Nature of the Human Genome," *APMIS* 124, nos. 1–2 (2016): 11–19, https://doi.org /10.1111/apm.12488; NIH, National Institute of General Medical

Sciences (NIGMS), "Our Complicated Relationship with Viruses," *ScienceDaily*, November 28, 2016, www.sciencedaily.com/releases/2016/11/161128151050.htm.

17. A. Prachumwat and W. H. Li, "Gene Number Expansion and Contraction in Vertebrate Genomes with Respect to Invertebrate Genomes," *Genome Research* 18, no. 2 (2008): 221–232.

18. L. Margulis and R. Fester, *Symbiosis as a Source of Evolutionary Innovation: Speciation and Morphogenesis* (Cambridge, MA: MIT Press, 1991); R. Guerrero, L. Margulis, and M. Berlanga, "Symbiogenesis: The Holobiont as a Unit of Evolution," *International Microbiology* 16, no. 3 (2013): 133–143.

19. E. Yong, *I Contain Multitudes: The Microbes within Us and a Grander View of Life* (New York: Penguin Random House, 2016).

20. N. A. Moran, "Accelerated Evolution and Muller's Rachet in Endosymbiotic Bacteria," *Proceedings of the National Academy of Sciences* 93, no. 7 (1996): 2873–2878; N. A. Moran, "Symbiosis as an Adaptive Process and Source of Phenotypic Complexity," *Proceedings of the National Academy of Sciences* 104, suppl. 1 (2007): 8627–8633.

21. M. McFall-Ngai, M. G., Hadfield, T. C. Bosch, H. V. Carey, T. Domazet-Lošo, A. E. Douglas, et al., "Animals in a Bacterial World, a New Imperative for the Life Sciences," *Proceedings of the National Academy of Sciences* 110, no. 9 (2013): 3229–3236.

22. M. Planck, *Scientific Autobiography, and Other Papers with a Memorial Address on Max Planck by Max von Laue*, trans. F. Gaynor (London: Williams & Norgate, 1950).

SPECIATION: THE FORMATION OF NEW SPECIES

1. S. P. Otto and J. Whitton, "Polyploid Incidence and Evolution," *Annual Review in Genetics* 34 (2000): 401–437; O. Paun, F. Forest, M. F. Fay, and M. W. Chase, "Hybrid Speciation in Angiosperms: Parental Divergence Drives Ploidy," *New Phytologist* 182, no. 2 (2009): 507–518.

2. A. K. Bergfeld, R. Lawrence, S. L. Diaz, O. M. T. Pearce, D. Ghaderi, P. Gagneux, et al., "*N*-Glycolyl Groups of Nonhuman Chondroitin Sulfates Survive in Ancient Fossils," *Proceedings of the National Academy of Sciences* 114, no. 39 (2017): E8155–E8164, https://doi.org/10.1073/pnas.1706306114.

3. C. Wilcox, "Extra DNA May Make Unlikely Hybrid Fish Possible," *Quanta Magazine*, August 5, 2020; J. Káldy, A. Mozsár, G. Fazekas, M. Farkas, D. L. Fazekas, G. L. Fazekas, et al., "Hybridization of Russian Sturgeon (*Acipenser gueldenstaedtii*, Brandt and Ratzeberg, 1833) and American Paddlefish (*Polyodon spathula*, Walbaum 1792) and Evaluation of Their Progeny," *Genes* 11, no. 7 (2020): 753.

4. J. Z. M. Chan, M. R. Halachev, N. J. Loman, C. Constantinidou, and M. J. Pallen, "Defining Bacterial Species in the Genomic Era: Insights from the Genus *Acinetobacter*," *BMC Microbiology* 12, article 302 (2012), https://doi.org/10.1186/1471-2180-12-302.

5. C. Darwin, *Journal of Researches into the Natural History and Geology of the Countries Visited during the Voyage of HMS Beagle round the World, under the Command of Capt. Fitz Roy* (London: Ward, Lock, 1889).

6. Darwin, *Journal of Researches*; J. Weiner, *The Beak of the Finch: A Story of Evolution in Our Time* (New York: Vintage Books/Random House, 1994); L. F. De León, D. M. Sharpe, K. M. Gotanda, J. A. Raeymaekers, J. A. Chaves, A. P. Hendry, and J. Podos, "Urbanization Erodes Niche Segregation in Darwin's Finches," *Evolutionary Applications* 12, no. 7 (2018): 1329–1343.

ON FOSSILS AND THE BOOK OF LIFE

1. R. Owen, "Darwin on the Origin of Species" (review published anonymously), *Edinburgh Review* 111, no. 8 (1860): 487–532; C. Darwin, C., *On the Origin of Species by Means of Natural Selection, Or the Preservation of Favoured Races in the Struggle for Life* (London: John Murray, 1861), 3.

2. Darwin, *On the Origin of Species* (1861 edition), 3; Stott, *Darwin's Ghosts*; S. M. Kidwell and K. W. Flessa, "The Quality of the Fossil Record: Populations, Species, and Communities," *Annual Review of Ecology and Systematics* 26 (1995): 269–299.

3. R. J. Baumgartner, M. J. Van Kranendonk, D. Wacey, M. L. Fiorentini, M. Saunders, S. Caruso, et al., "Nano-Porous Pyrite and Organic Matter in 3.5-billion-Year-Old Stromatolites Record Primordial Life," *Geology* 47, no. 11 (2019): 1039–1043, https://doi.org/10.1130/G46365.1; M. S. Dodd, D. Papineau, T. Grenne, J. F. Slack, M. Rittner, F. Pirajno, et al., "Evidence for Early Life in Earth's Oldest

Hydrothermal Vent Precipitates," *Nature* 543, no. 7643 (2017): 60–64, https://doi.org/10.1038/nature21377.

4. S. E. Peters, "Environmental Determinants of Extinction Selectivity in the Fossil Record," *Nature* 454 (2008): 626–629, https://doi.org/10.1038/nature07032.

5. J. Alroy, "How Many Named Species Are Valid?" *Proceedings of the National Academy of Sciences* 99, no. 6 (2002): 3706–3711, https://doi.org/10.1073/pnas.062691099.

6. P. Williams, *The Dinosaur Artist Obsession, Betrayal, and the Quest for Earth's Ultimate Trophy* (New York: Hachette Books, 2018).

7. T. Appel, *The Cuvier-Geoffroy Debate: French Biology in the Decades before Darwin* (New York: Oxford University Press, 1987).

8. Appel, *Cuvier-Geoffroy Debate*; E. Lurie, "Louis Agassiz and the Idea of Evolution," *Victorian Studies* (Indiana University Press) 3, no. 1 (1959): 87–108, http://www.jstor.org/stable/3825589.

9. L. Miller, *Why Fish Don't Exist* (New York: Simon & Schuster, 2020); B. Forrest and P. R. Gross, *Creationism's Trojan Horse: The Wedge of Intelligent Design* (New York: Oxford University Press, 2004), https://doi.org/10.1093/acprof:oso/9780195157420.001.0001.

10. P. D. Brinkman, "Charles Darwin's *Beagle* Voyage, Fossil Vertebrate Succession, and 'The Gradual Birth & Death of Species,'" *Journal of the History of Biology* 43 (2009): 363–399, https://doi.org/10.1007/s10739-009-9189-9.

11. Brinkman, "Charles Darwin's *Beagle* Voyage"; C. Darwin, *Voyages of the Adventure and Beagle,* vol. III, *Journal and Remarks* (London: Henry Colburn, 1839).

12. C. Lyell, *Principles of Geology: Being an Attempt to Explain the Former Changes of the Earth's Surface by Reference to Causes Now in Operation* (London: John Murray, 1830–1833), vol. 1.

13. G. Gohau, "Darwin the Geologist: Between Lyell and von Buch" ["Darwin le géologue: Entre Lyell et von Buch"], *Comptes Rendus Biologies* 333, no. 2 (2010): 95–98, https://doi.org/10.1016/j.crvi.2009.11.008.

14. Darwin, *On the Origin of Species* (1861 edition), 3.

15. P. Wellnhofer, "A Short History of Research on *Archaeopteryx* and Its Relationship with Dinosaurs," *Geological Society, Special Publications* 343, no. 1 (2010): 237–250, https://doi.org/10.1144/SP343.14.

16. Owen, "Darwin on the Origin of Species."

17. Darwin, *On the Origin of Species* (1861 edition), 3.

18. Stott, *Darwin's Ghosts.*

19. Wellnhofer, "Short History of Research on *Archaeopteryx*"; P. D. Gingerich, "Land-to-Sea Transition in Early Whales: Evolution of Eocene Archaeoceti (Cetacea) in Relation to Skeletal Proportions and Locomotion of Living Semiaquatic Mammals, *Paleobiology* 29, no. 3 (2003): 429–454, https://doi.org/10.1666/0094-8373(2003)029<0429 :LTIEWE>2.0.CO;2; I. Schneider and N. H. Shubin, "The Origin of the Tetrapod Limb: From Expeditions to Enhancers," *Trends in Genetics* 29, no. 7 (2013): 419–426, https://doi.org/10.1016/j.tig.2013.01.012; T. A. Stewart, J. B. Lemberg, A. Daly, E. B. Daeschler, and N. H. Shubin, "A New Elpistostegalian from the Late Devonian of the Canadian Arctic," *Nature* 603 (2022): 563–563.

20. Gingerich, "Land-to-Sea Transition in Early Whales."

21. I. Schneider and N. H. Shubin, "The Origin of the Tetrapod Limb: From Expeditions to Enhancers," *Trends in Genetics* 29, no. 7 (2013): 419–426, https://doi.org/10.1016/j.tig.2013.01.012.

22. Stewart et al., "New Elpistostegalian."

23. T. R. Lyson, G. S. Bever, B. A. S. Bhullar, W. G. Joyce, and J. A. Gauthier, "Transitional Fossils and the Origin of Turtles," *Paleontology* 6 (2010): 830–833, https://doi.org/10.1098/rsbl.2010.0371.

24. C. B. String and P. Andrews, "Genetic and Fossil Evidence for the Origin of Modern Humans," *Science* 239, no. 4845 (1988): 1263–1268, https://doi.org/10.1126/science.3125610.

PART III

WHO YOU CALLING "PRIMITIVE"?

1. C. H. Tyndale-Biscoe, *The Life of Marsupials* (Clayton, Australia: CSIRO, 2005).

2. H. M. Dunsworth, "There is No 'Obstetrical Dilemma': Toward a Braver Medicine with Fewer Childbirth Interventions," *Perspectives in Biology and Medicine* 61, no. 2 (2018): 249–263.

3. N. S. Upham, J. A. Esselstyn, and W. Jetz, "Inferring the Mammal Tree: Species-Level Sets of Phylogenies for Questions in

Ecology, Evolution, and Conservation," *PLOS Biology* 17, no. 12 (2019): e3000494, https://doi.org/10.1371/journal.pbio.3000494.

4. T. Grant, *Platypus*, 4th ed. (Clayton, Australia: CSIRO, 2007).

5. P. S. Anich, S. Anthony, M. Carlson, A. Gunnelson, A. M. Kohler, J. G. Martin, and E. R. Olson, "Biofluorescence in the Platypus (*Ornithorhynchus anatinus*)," *Mammalia* 85, no. 2 (2020): 179–181.

6. L. Spinney, "Remnants of Evolution," *New Scientist* 198, no. 2656 (2008): 42–45.

7. Shubin, *Your Inner Fish*.

8. J. A. Clack and P. E. Ahlberg, "Sarcopterygians: From Lobe-Ffinned Fishes to the Tetrapod Stem Group," in *Evolution of the Vertebrate Ear: Evidence from the Fossil Record*, ed. J. A. Clack, R. R. Fay, and A. N. Popper (New York: Springer, 2016), 51–70, Springer Handbook of Auditory Research series, vol. 59.

9. R. Cloutier, A. M. Clement, M. S. Y. Lee, R. Noël, I. Béchard, V. Roy, and J. A. Long, "*Elpistostege* and the Origin of the Vertebrate Hand," *Nature* 579 (2020): 549–554, https://doi.org/10.1038/s41586 -020-2100-8.

10. I. Braasch, A. R. Gehrke, J. J. Smith, K. Kawasaki, T. Masousaki, J. Pasquier, et al., "The Spotted Gar Genome Illuminates Vertebrate Evolution and Human-Teleost Comparisons," *Nature Genetics* 48 (2016): 427–437, https://doi.org/10.1038/ng.3526.

11. Braasch et al., "Spotted Gar Genome."

OUR CRAPPY-CRAPPIE BODIES

1. S. Olshansky, B. Carnes, and R. Butler, "If Humans Were Built to Last," *Scientific American* 284, no. 3 (March 2001): 95–100.

2. S. Giles, G. Xu, T. J. Near, and M. Friedman, "Early Members of 'Living Fossil' Lineage Imply Later Origin of Modern Ray-Finned Fishes," *Nature* 549 (2017): 265–268, https://doi.org/10.1038/nature 23654; M. L. Stiassny, L. R.Parenti, and G. D. Johnson, eds., *Interrelationships of Fishes* (San Diego, CA: Academic Press, 1996); L. C. Hughes, G. Ortí, Y. Huang, Y. Sun, C. C. Baldwin, A. W. Thompson, et al., "Comprehensive Phylogeny of Ray-Finned Fishes (Actinopterygii) Based on Transcriptomic and Genomic Data," *Proceedings of the National Academy of Sciences* 115, no. 24 (2018): 6249–6254.

3. G. S. Helfman, B. B. Collette, and D. E. Facey, *The Diversity of Fishes* (Malden, MA: Blackwell Science, 1997).

4. . M. Gemberling, T. J. Bailey, D. R. Hyde, and K. D. Poss, "The Zebrafish as a Model for Complex Tissue Regeneration," *Trends in Genetics* 29, no. 11 (2013): 611–620; Williams, R., "Thanks Be to Zebrafish," *Circulation Research* 107, no. 5 (2010): 570–572.

5. R. Diogo, V. Abdala, N. Lonergan, and B. A. Wood, "From Fish to Modern Humans—Comparativ Anatomy, Homologies and Evolution of the Head and Neck Musculature," *Journal of Anatomy* 213, no. 4 (2008): 391–424, https://doi.org/10.1111/j.1469-7580.2008.00953.x.

6. C. A. Leite, E. W. Taylor, C. D. Guerra, L. H. Florindo, T. Belão, and F. T. Rantin, "The Role of the Vagus Nerve in the Generation of Cardiorespiratory Interactions in a Neotropical Fish, the Pacu, *Piaractus mesopotamicus*," *Journal of Comparative Physiology A* 195, no. 8 (2009): 721–731, https://doi.org/10.1007/s00359-009-0447-2.

7. M. J. Wedel, "A Monument of Inefficiency: The Presumed Course of the Recurrent Laryngeal Nerve in Sauropod Dinosaurs," *Acta Palaeontologica Polonica* 57, no. 2 (2011): 251–256.

8. Leite et al., "The Role of the Vagus Nerve"; B. Bryson, *The Body: A Guide for Occupants* (Doubleday/Penguin Random House, 2019); T. Nishimura, I. T. Tokuda, S. Miyachi, J. C. Dunn, C. T. Herbst, K. Ishimura, K., et al., "Evolutionary Loss of Complexity in Human Vocal Anatomy as an Adaptation for Speech," *Science* 377, no. 6607 (2022): 760–763.

9. Olshansky, Carnes, and Butler, "If Humans Were Built to Last."

THE TREE OF LIFE

1. W. N. Eschmeyer and California Academy of Sciences, "Eschmeyer's Catalog of Fishes: Genera, Species, References," open-access online version, last updated October 4, 2022, http://researcharchive .calacademy.org/research/ichthyology/catalog/fishcatmain.asp.

2. N. E. Stork, J. McBroom, C. Gely, and A. J. Hamilton, "New Approaches to Narrow Global Species Estimates for Beetles, Insects, and Terrestrial Arthropods," *Proceedings of the National Academy of Sciences* 112, no. 24 (2015): 7519–7523, https://www.pnas.org/content /pnas/112/24/7519.full.pdf; D. Dykhuizen, "Species Numbers in

Bacteria," *Proceedings of the California Academy of Science* 56, no. 6, suppl. 1 (2011): 62–71, https://www.ncbi.nlm.nih.gov/pmc/articles/PMC 3160642/.

3. "Ocean Metagenomics in PLOS Biology: A collection of articles from the J. Craig Venter Institute's Global Ocean Sampling Expedition." See, in particular, S. Yooseph, G. Sutton, D. B. Rusch, A. L. Halpern, AL; S. J. Williamson, K. Remington, et al., "The Sorcerer II Global Ocean Sampling Expedition: Expanding the Universe of Protein Families, *PLOS Biology* 5, no. 3 (2007): e16, https://doi.org/10 .1371/journal.pbio.0050016; N. Kannan, S. S. Taylor, Y. Zhai, J. C. Venter, and G. Manning, "Structural and Functional Diversity of the Microbial Kinome," *PLOS Biology* 5, no. 3 (2007): e17. https://doi.org /10.1371/journal.pbio.005001; D. B. Rusch, A. L. Halpern, G. Sutton, K. B. Heidelberg, S. Williamson, S. Yooseph, et al., "The Sorcerer II Global Ocean Sampling Expedition: Northwest Atlantic through Eastern Tropical Pacific," *PLOS Biology* 5, no. 3 (2007): e77, https://doi .org/10.1371/journal.pbio.0050077; N. Yutin, M. T. Suzuki, H. Teeling, M. Weber, J. C. Venter, D. B. Rusch, and O. Béjà, "Assessing Diversity and Biogeography of Aerobic Anoxygenic Phototrophic Bacteria in Surface Waters of the Atlantic and Pacific Oceans using the Global Ocean Sampling Expedition Metagenomes," *Environmental Microbiology* 9, no. 6 (2007): 1464–1475, https://doi.org/10.1111/j .1462-2920.2007.01265.x; and I. Sharon, S. Tzahor, S. Williamson, M. Shmoish, D. Man-Aharonovich, D. B. Rusch, et al., "Viral Photosynthetic Reaction Center Genes and Transcripts in the Marine Environment," *ISME Journal* 1, no. 6 (2007): 492–501, https://doi.org/10.1038 /ismej.2007.67, listed at "Global Ocean Sampling Expedition," Wikipedia, last edited May 30, 2022, https://en.wikipedia.org/wiki/Global _Ocean_Sampling_Expedition.

4. "Ocean Metagenomics in PLOS Biology."

5. P. Chakrabarty, "Four Billion Years of Evolution in Six Minutes," TED.com, April 2018, https://www.ted.com/talks/prosanta_chakrab arty_four_billion_years_of_evolution_in_six_minutes?language=en.

6. "Open Tree of Life" (illustration), *One Zoom*, viewed October 7, 2022, http://www.onezoom.org/life.html.

7. M. J. Benton and R. J. Twitchett, "How to Kill (Almost) All Life: The End-Permian Extinction Event," *Trends in Ecology & Evolution* 18,

no. 7 (2003): 358–365, https://www.sciencedirect.com/science/article/abs/pii/S0169534703000934.

8. R. M. May and D. L. Hawksworth, "Conceptual Aspects of the Quantification of the Extent of Biological Diversity," *Philosophical Transactions of the Royal Society* 345, no. 1311 (1994): 13–20, http://doi.org/10.1098/rstb.1994.0082.

9. I. Mukherjee, R. R. Large, R. Corkrey, and L. V. Danyushevsky, "The Boring Billion, a Slingshot for Complex Life on Earth," *Scientific Reports* 8, no. 4432 (2018), https://doi.org/10.1038/s41598-018-22695-x.

10. C. R. Woese, O. Kandler, and M. L. Wheelis, "Towards a Natural System of Organisms: Proposal for the Domains Archaea, Bacteria, and Eucarya," *Proceedings of the National Academy of Sciences* 87, no. 12 (1990): 4576–4579, https://doi.org/10.1073/pnas.87.12.4576.

11. L. Margulis, "Symbiogenesis: A New Principle of Evolution Rediscovery of Boris Mikhaylovich Kozo-Polyansky (1890–1957)," *Paleontological Journal* 44, no. 12 (2011): 1525–1539, https://doi.org/10.1134/S0031030110120087.

12. J. A. Theriot, "Why Are Bacteria Different from Eukaryotes?' *BMC Biology* 11, article 119 (2013), https://doi.org/10.1186/1741-7007-11-119.

13. Theriot, "Why Are Bacteria Different?"

14. *Encyclopaedia Britannica online*, q.v. "category" (logic), accessed October 7, 2022, https://www.britannica.com/topic/category-logic; Aristotle, *De Anima (On the Soul)*, trans. H. Lawson-Tancred (Harmondsworth: Penguin, 1986).

15. C. M. Cavanaugh, S. L. Gardiner, M. L. Jones, H. W. Jannasch, and J. B. Waterbury, "Prokaryotic Cells in the Hydrothermal Vent Tube Worm *Riftia pachyptila* Jones: Possible Chemoautotrophic Symbionts," *Science* 213, no. 4505 (1981): 340–342.

THE ORIGIN OF LIFE

1. D. Macdougall, *Endless Novelties of Extraordinary Interest: The Voyage of H.M.S. Challenger and the Birth of Modern Oceanography* (New Haven: Yale University Press, 2019).

2. C. Zimmer, *Life's Edge: The Search for What It Means to Be Alive* (London: Pan Macmillan, 2021).

3. E. Sober, *Evidence and Evolution: The Logic behind the Science* (Cambridge: Cambridge University Press, 2008).

4. K. Popper, *Conjectures and Refutations: The Growth of Scientific Knowledge* (New York: Basic Books, 1962); D. Deutsch, *The Beginning of Infinity: Explanations That Transform the World* (Harmondsworth: Penguin, 2011).

5. J. A. G. Ranea, A. Sillero, J. M. Thornton, and C. A. Orengo, "Protein Superfamily Evolution and the Last Universal Common Ancestor (LUCA)," *Journal of Molecular Evolution* 63, no. 4 (2006): 513–525, https://doi.org/10.1007/s00239-005-0289-7.

6. N. Wolchover, "First Support for a Physics Theory of Life," *Quanta* magazine, July 26, 2017, https://www.quantamagazine.org/first-support-for-a-physics-theory-of-life-20170726/.

7. S. L. Miller and H. C. Urey, "Organic Compound Synthesis on the Primitive Earth," *Science* 130, no. 3370 (1959): 245–251.

8. K. Than, "Organic Molecules Found in Diverse Space Places," Space.com, August 8, 2006, http://www.space.com/2711-organic-molecules-diverse-space-places.html; D. Deamer, *First Life: Discovering the Connections between Stars, Cells, and How Life Began* (Berkeley: University of California Press, 2011).

9. L. I. Cleeves, E. A. Bergin, C. M. D. Alexander, F. Du, D. Graninger, K. I. Öberg, and T. J. Harries, "The Ancient Heritage of Water Ice in the Solar System," *Science* 345, no. 6204 (2014): 1590–1593.

10. J. Zalasiewicz and M. Williams, *Ocean Worlds: The Story of Seas on Earth and Other Planets* (New York: Oxford University Press, 2014).

11. J. S. Greaves, A. M. Richards, W. Bains, P. B. Rimmer, H. Sagawa, D. L. Clements, et al., "Phosphine Gas in the Cloud Decks of Venus," *Nature Astronomy* 5 (2020): 655–664.

12. W. Bains, J. J. Petkowski, S. Seager, S. Ranjan, C. Sousa-Silva, P. B. Rimmer, et al., "Phosphine on Venus Cannot Be Explained by Conventional Processes," *Astrobiology* 21, no. 10 (2021): 1277–1304.

13. G. Wald, "The Origin of Optical Activity," *Annals of the New York Academy of Sciences* 69, no. 2 (1957): 352–368, https://doi.org/10.1111/j.1749-6632.1957.tb49671.x.

14. J. Bailey, "Chirality and the Origin of Life," *Acta Astronautica* 46, nos. 10–12 (2000): 627–631.

15. T. D. Brock, "The Value of Basic Research: Discovery of *Thermus aquaticus* and Other Extreme Thermophiles," *Genetics* 146, no. 4 (1997): 1207–1210.

16. Baumgartner et al., "Nano-Porous Pyrite and Organic Matter."

17. Baumgartner et al., "Nano-Porous Pyrite and Organic Matter."

THE HISTORY OF LIFE

1. S. Xiao and Q. Tang, "After the Boring Billion and before the Breezing Billion: Evolutionary Patterns and Innovations in the Tonian Period," *Emerging Topics in Life Sciences* 2, no. 2 (2018): 161–171, https://doi.org/10.1042/ETLS20170165.

2. M. J. McFall-Ngai, "Giving Microbes Their Due—Animal Life in a Microbially Dominant World," *Journal of Experimental Biology* 218, no. 12 (2015): 1968–1973, https://doi.org/10.1242/jeb.115121.

3. S. M. Hird, "Evolutionary Biology Needs Wild Microbiomes," *Frontiers in Microbiology* 8, no. 725 (2017), https://www.frontiersin.org/article/10.3389/fmicb.2017.00725; L. Dethlefsen, M. McFall-Ngai, and D. Relman, "An Ecological and Evolutionary Perspective on Human-Microbe Mutualism and Disease," *Nature* 449 (2007): 811–818, https://doi.org/10.1038/nature06245.

4. McFall-Ngai, "Giving Microbes Their Due."

5. T. Lyons, C. Reinhard, and N. Planavsky, "The Rise of Oxygen in Earth's Early Ocean and Atmosphere," *Nature* 506 (2014): 307–315, https://doi.org/10.1038/nature13068; K. J. Zahnle, D. C. Catling, and M. W. Claire, "The Rise of Oxygen and the Hydrogen Hourglass," *Chemical Geology* 362 (2013): 26–34, https://doi.org/10.1016/j.chemgeo.2013.08.004.

6. M. S. W. Hodgskiss, P. W. Crockford, Y. Peng, B. A. Wing, and T. J. Horner, "A Productivity Collapse to End Earth's Great Oxidation," *Proceedings of the National Academy of Sciences* 116, no. 35 (2019): 17207–17212, https://doi.org/10.1073/pnas.1900325116.

7. T. Hemnani and M. S. Parihar, "Reactive Oxygen Species and Oxidative DNA Damage," *Indian Journal of Physiology and Pharmacology* 42, no. 4 (1998): 440–452.

8. X. Chen, H. Ling, D. Vance, G. A. Shields-Zhou, M. Zhu, S. W. Poulton, et al., "Rise to Modern Levels of Ocean Oxygenation Coincided

with the Cambrian Radiation of Animals," *Nature Communications* 6, article 7142 (2015), https://doi.org/10.1038/ncomms8142.

9. A. H. Knoll, "The Early Evolution of Eukaryotes: A Geological Perspective," *Science* 256, no. 5057 (1992): 622–627; B. Runnegar, "Loophole for Snowball Earth," *Nature* 405 (2000): 403–404, https://doi.org/10.1038/35013168; A. Rokas, "The Molecular Origins of Multicellular Transitions," *Current Opinion in Genetics & Development* 18, no. 6 (2008): 472–478, https://doi.org/10.1016/j.gde.2008.09.004; N. King, "The Unicellular Ancestry of Animal Development," *Developmental Cell* 7, no. 3 (2004): 313–325, https://doi.org/10.1016/j.devcel.2004.08.010.

10. Knoll, "The Early Evolution of Eukaryotes."

11. B. Runnegar, "Loophole for Snowball Earth," *Nature* 405 (2000): 403–404, https://doi.org/10.1038/35013168.

12. S. Morris and J. Caron, "A Primitive Fish from the Cambrian of North America," *Nature* (2014): 419–422, https://doi.org/10.1038/nature13414.

13. S. C. Morris, "Burgess Shale Faunas and the Cambrian Explosion," *Science* 246, no. 4928 (1989): 339–346.

14. C. C. Loron, C. François, R. H. Rainbird, R. H., E. C. Turner, S. Borensztajn, and E. J. Javaux, "Early Fungi from the Proterozoic Era in Arctic Canada," *Nature* 570 (2019): 232–235, https://doi.org/10.1038/s41586-019-1217-0; D. S. Heckman, D. M. Geiser, B. R. Eidell, R. L. Stauffer, N. L. Kardos, and S. B. Hedges, "Molecular Evidence for the Early Colonization of Land by Fungi and Plants," *Science* 293, no. 5532 (2001): 1129–1133; R. K. Grosberg, G. J. Vermeij, and P. C. Wainwright, "Biodiversity in Water and on Land," *Current Biology* 22, no. 21 (2012): 900–903.

15. Loron et al., "Early Fungi from the Proterozoic Era."

16. Heckman et al., "Molecular Evidence for the Early Colonization of Land."

17. Z. Luo, "Transformation and Diversification in Early Mammal Evolution," *Nature* 450 (2007): 1011–1019, https://doi.org/10.1038/nature06277; C. M. Janis, "Tertiary Mammal Evolution in the Context of Changing Climates, Vegetation, and Tectonic Events," *Annual Review of Earth and Planetary Sciences* 24 (1993): 467–500; D. E. Wilson and D. M. Reeder, *Mammal Species of the World: A Taxonomic*

and Geographic Reference (Baltimore: Johns Hopkins University Press, 2005).

18. J. Hawks and L. Berger, "*Homo naledi* and Its Significance in Evolutionary Anthropology," in *Theology and Evolutionary Anthropology: Dialogues in Wisdom, Humility and Grace*, ed. C. D. Drummond and A. Fuentes (London: Routledge, 2020), 51–68; P. Andrews and T. Harrison, "The Last Common Ancestor of Apes and Humans," in *Interpreting the Past: Essays on Human, Primate, and Mammal Evolution in Honor of David Pilbeam*, ed. J. Kelley, D. E. Lieberman, and R. J. Smith (Boston: Brill Academic, 2005), 103–121.

19. Y. M. Bar-On, R. Phillips, and R. Milo, "The Biomass Distribution on Earth," *Proceedings of the National Academy of Sciences* 115, no. 25 (2018): 6506–6511, https://doi.org/10.1073/pnas.1711842115.

20. Bar-On, Phillips, and Milo, "The Biomass Distribution on Earth."

PART IV

EVOLUTION IN THE ANTHROPOCENE

1. B. Van Valkenburgh, "Déjà Vu: The Evolution of Feeding Morphologies in the Carnivora," *Integrative & Comparative Biology* 47, no. 1 (2007): 147–163.

2. D. Sargan, "Ethics and Genetics of Brachycephalic Breeding: Can They All Be Bad?" in *British Small Animal Veterinary Association Congress 2018 Proceedings*, 370–371, https://doi.org/10.22233/978191044 3590.50.1.

3. J. Wang, W. G. Hill, D. Charlesworth, and B. Charlesworth, "Dynamics of Inbreeding Depression Dtaskbarue to Deleterious Mutations in Small Populations: Mutation Parameters and Inbreeding Rate," *Genetics Research* 74, no. 2 (1999): 165–178; A. P. Wolf, *Inbreeding, Incest, and the Incest Taboo: The State of Knowledge at the Turn of the Century* (Stanford: Stanford University Press, 2005).

4. F. Jiang and J. A. Doudna, "CRISPR-Cas9 Structures and Mechanisms," *Annual Review of Biophysics* 46, no. 1 (2017): 505–529.

5. A. Buchman, S. Gamez, M. Li, I. Antoshechkin, H. Li, H. Wang, et al., "Engineered Resistance to Zika Virus in Transgenic *Aedes aegypti* Expressing a Polycistronic Cluster of Synthetic Small RNAs," *Proceedings of the National Academy of Sciences* 116, no. 9 (2019):

3656–3661, https://doi.org/10.1073/pnas.1810771116; M. Scudellari, "Self-Destructing Mosquitoes and Sterilized Rodents: The Promise of Gene Drives," *Nature* 571 (2019): 160–162, https://doi.org/10.1038/d41586-019-02087-5.

6. K. S. Bosley, M. Botchan, A. L. Bredenoord, D. Carroll, R. A. Charo, E. Charpentier, et al., "CRISPR Germline Engineering—The Community Speaks," *Nature Biotechnology* 33, no. 5 (2015): 478–486; D. Baltimore, P. Berg, M. Botchan, D. Carroll, R. A. Charo, G. Church, et al., "A Prudent Path Forward for Genomic Engineering and Germline Gene Modification," *Science* 348, no. 6230 (2015): 36–38; J. Doudna, "Genome-Editing Revolution: My Whirlwind Year with CRISPR," *Nature News* 528, no. 7583 (2015): 469.

7. S. Tiwari, N. Kaur, and P. Awasthi, "Fruit Crops Improvement Using CRISPR/Cas9 System," in *Genome Engineering via CRISPR-Cas9 System*, ed. V. Singh and P. K. Dhar (London: Academic Press, 2020), 131–145, https://doi.org/10.1016/b978-0-12-818140-9.00012-x.

8. T. G. Knowles, S. C. Kestin, S. M. Haslam, S. N. Brown, L. E. Green, A. Butterworth, et al., "Leg Disorders in Broiler Chickens: Prevalence, Risk Factors and Prevention," *PLOS ONE* 3, no. 2 (2008): e1545, https://doi.org/10.1371/journal.pone.0001545; W. Isaacson, *The Code Breaker: Jennifer Doudna, Gene Editing, and the Future of the Human Race* (New York: Simon & Schuster, 2021).

9. Buchman et al., "Engineered Resistance to Zika Virus"; Scudellari, "Self-Destructing Mosquitoes and Sterilized Rodents."

10. R. I. Taitingfong, "Islands as Laboratories: Indigenous Knowledge and Gene Drives in the Pacific," *Human Biology* 91, no. 3 (2020): 179–188, Project MUSE, https://muse.jhu.edu/article/749634/summary.

11. M. C. Wong, S. J. J. Cregeen, N. J. Ajami, and J. F. Petrosino, "Evidence of Recombination in Coronaviruses Implicating Pangolin Origins of nCoV-2019," *bioRxiv*, February 13, 2020, https://doi.org/10.1101/2020.02.07.939207; K. G. Andersen, A. Rambaut, W. I. Lipkin, E. C. Holmes, and R. F. Garry, "The Proximal Origin of SARS-CoV-2," *Nature Medicine* 26 (2020): 450–452, https://doi.org/10.1038/s41591-020-0820-9.

12. Andersen et al., "Proximal Origin of SARS-CoV-2."

13. D. Simberloff, D. C. Schmitz, and T. C. Brown, *Strangers in Paradise: Impact and Management of Nonindigenous Species in Florida* (Washington, DC: Island Press, 1997).

14. D. Adhikari, R. Tiwary, and S. K. Barik, "Modelling Hotspots for Invasive Alien Plants in India," *PLOS ONE* 10, no. 7 (2015): e0134665; A. J. Ankila Hiremath and S. Krishnan, "India Knows Its Invasive Species Problem but This Is Why Nobody Can Deal with It Properly," *Wire*, December 11, 2016.

15. Zimmer, *She Has Her Mother's Laugh*; E. Kleiderman and U. Ogbogu, "Realigning Gene Editing with Clinical Research Ethics: What the 'CRISPR Twins' Debacle Means for Chinese and International Research Ehics Governance," *Accountability in Research* 26, no. 4 (2019): 257–264, https://doi.org/10.1080/08989621.2019.1617138.

16. A. Park, "How Jennifer Doudna's Life Has Changed since Discovering CRISPR 10 Years Ago," *Time*, July 1, 2022, https://time.com/6192962/crispr-jennifer-doudna-10-year-anniversary/.

17. Y. N. Harari, *Homo Deus: A Brief History of Tomorrow* (London: Harvill Secker, 2015).

18. D. F. Maron, "Under Poaching Pressure, Elephants Are Evolving to Lose Their Tusks," National Geographic.com, November 9, 2018, https://www.nationalgeographic.com/animals/2018/11/wildlife-watch-news-tuskless-elephants-behavior-change/; S. C. Campbell-Staton, B. J. Arnold, D. Gonçalves, P. Granli, J. Poole, R. A. Long, and R. M. Pringle, "Ivory Poaching and the Rapid Evolution of Tusklessness in African Elephants," *Science* 374, no. 6566 (2021): 483–487.

19. R. Ellis, "The Bluefin in Peril," *Scientific American* 298, no. 3 (March 2008): 70–77, https://www.jstor.org/stable/26000519.

20. R. Piper, *Extinct Animals: An Encyclopedia of Species That Have Disappeared during Human History* (Westport, CT: Greenwood Press, 2009).

21. R. E. Leakey and R. Lewin, *The Sixth Extinction: Patterns of Life and the Future of Humankind* (New York: Anchor/Books/Random House, 1996); G. Ceballos, P. R. Ehrlich, A. D. Barnosky, A. García, R. M. Pringle, and T. M. Palmer, "Accelerated Modern Human-Induced Species Losses: Entering the Sixth Mass Extinction," *Science Advances* 1, no. 5 (2015): e1400253; S. T. Turvey and J. J. Crees, "Extinction in the Anthropocene," *Current Biology* 29, no. 19 (2019): R982–R986; S. Pimm and P. Raven, "Extinction by Numbers," *Nature* 403 (2000): 843–845, https://doi.org/10.1038/35002708; E. Kolbert, *The Sixth Extinction: An Unnatural History* (London: A&C Black, 2014).

22. Pimm and Raven, "Extinction by Numbers"; Kolbert, *Sixth Extinction*; T. M. Brooks, R. A. Mittermeier, C. G. Mittermeier, G. A. B. da Fonseca, A. B. Rylands, W. R. Konstant, et al., "Habitat Loss and Extinction in the Hotspots of Biodiversity," *Conservation Biology* 16, no. 4 (2002): 909–923, https://doi.org/10.1046/j.1523-1739.2002.00530.x.

23. K. M. Gaynor, C. E. Hojnowski, N. H. Carter, and J. S. Brashares, "The Influence of Human Disturbance on Wildlife Nocturnality," *Science* 360, no. 6394 (2018): 1232–1235.

24. P. J. Crutzen, "The 'Anthropocene,'" in *Earth System Science in the Anthropocene* (Berlin: Springer, 2006), 13–18.

25. S. Osaka, "The World Is on Lockdown: So Where Are All the Carbon Emissions Coming From?" *Grist*, April 27, 2020, https://grist .org/climate/the-world-is-on-lockdown-so-where-are-all-the-carbon -emissions-coming-from/.

NATURAL HISTORY

1. V. Bush, "Science, the Endless Frontier: A Report to the President" (Washington, DC: US Government Printing Office, 1945), https:// www.nsf.gov/about/history/vbush1945.htm; E. Morris, *Edison* (New York: Penguin Random House, 2019).

2. L. Rocha, A. Aleixo, F. Allen, C. C. Almeda, M. V. I. Baldwin, J. M. Barclay, et al., "Specimen Collection: An Essential Tool," *Science* 344, no. 6186 (2014): 814–815.

3. J. Lendemer, B. Thiers, A. K. Monfils, J. Zaspel, E. R. Ellwood, A. Bentley, et al., "The Extended Specimen Network: A Strategy to Enhance U.S.Biodiversity Collections, Promote Research and Education," *BioScience* 70, no. 1 (2020): 23–30; J. C. Buckner, R. C. Sanders, B. C. Faircloth, and P. Chakrabarty, "Science Forum: The Critical Importance of Vouchers in Genomics," *Elife* 10 (2021): e68264.

4. Z. A. Goodwin, D. J. Harris, D. Filer, J. R. Wood, and R. W. Scotland, "Widespread Mistaken Identity in Tropical Plant Collections," *Current Biology* 25, no. 22 (2015): R1066–R1067.

5. E. Larsson, "Collecting, Curating and the Construction of Zoological Knowledge: Walter Rothschild's Zoological Enterprise, c. 1878–1937," Ph.D. thesis, King's College London, 2019; C. Kemp, "The Financier Who Bought All of Nature," *New Scientist* 236, no. 3150 (2017): 40–41.

6. K. W. Johnson, *The Feather Thief: Beauty, Obsession, and the Natural History Heist of the Century* (New York: Penguin Random House, 2018).

7. P. Chakrabarty, M. P. Davis, W. L. Smith, Z. H. Baldwin, and J. S. Sparks, "Is Sexual Selection Driving Diversification of the Bioluminescent Ponyfishes (Teleostei: Leiognathidae)?" *Molecular Ecology* 20, no. 13 (2011): 2818–2834.

8. Chakrabarty et al., "Is Sexual Selection Driving Diversification?"

9. T. Cashion, F. Le Manach, D. Zeller, and D. Pauly, "Most Fish Destined for Fishmeal Production Are Food-Grade Fish," *Fish and Fisheries* 18, no. 5 (2017): 837–844, https://doi.org/10.1111/faf.12209.

10. K. Kauhala, K. Talvitie, and T. Vuorisalo, "Free-Ranging House Cats in Urban and Rural Areas in the North: Useful Rodent Killers or Harmful Bird Predators?" *Journal of Vertebrate Biology* 64, no. 1 (2015): 45–55; S. R. Loss, T. Will, and P. P. Marra, "The Impact of Free-Ranging Domestic Cats on Wildlife of the United States," *Nature Communications* 4, article 1396 (2013), http://dx.doi.org/10.1038/ncomms2380.

11. A. Mood and P. Brooke, "Estimating the Number of Fish Caught in Global Fishing Each Year," Fishcount.org.uk, July 2010, http://fishcount.org.uk/published/std/fishcountstudy.pdf.

12. Rocha et al., "Specimen Collection."

13. K. Wong, "Mother Nature's Medicine Cabinet," *Scientific American*, posted April 9, 2001. https://www.scientificamerican.com/article/mother-natures-medicine-c/.

14. D. DiEuliis, K. R. Johnson, S. S. Morse, and D. E. Schindel, "Opinion: Specimen Collections Should Have a Much Bigger Role in Infectious Disease Research and Response," *Proceedings of the National Academy of Sciences* 113, no. 1 (2011): 4–7, https://doi.org/10.1073/pnas.1522680112; C. W. Thompson, K. L. Phelps, M. W. Allard, J. A. Cook, J. L. Dunnum, A. W. Ferguson, et al., "Preserve a Voucher Specimen! The Critical Need for Integrating Natural History Collections in Infectious Disease Studies," *Mbio* 12, no. 1 (2021): e02698-20.

15. S. Warny, S. Ferguson, M. S. Hafner, and G. Escarguel, "Using Museum Pelt Collections to Generate Pollen Prints from High-Risk Regions: A New Palynological Forensic Strategy for Geolocation," *Forensic Science International* 306, no. 110061 (2019), https://doi.org/10.1016/j.forsciint.2019.110061.

16. J. Drew, "The Role of Natural History Institutions and Bioinformatics in Conservation Biology," *Conservation Biology* 25, no. 6 (2011):1250–1252, https://doi.org/10.1111/j.1523-1739.2011.01725.x.

17. S. Das and M. Lowe, "Nature Read in Black and White: Decolonial Approaches to Interpreting Natural History Collections," *Journal of Natural Science Collections* 6 (2018): 4–14; A. de Vos, "The Problem of 'Colonial Science,'" *Scientific American*, posted online July 1, 2020. https://www.scientificamerican.com/article/the-problem-of-colonial-science/; J. Ndunguru, F. Tairo, L. Boykin, and P. Sseruwagi, "Principles of Effective Collaboration in Agricultural Development and Research for Impact," *AgriRxiv*, preprint February 1, 2019, https://doi.org/10.31220/osf.io/udf63.

18. B. Fontaine, K. van Achterberg, M. A. Alonso-Zarazaga, R. Araujo, M. Asche, H. Aspöck, et al., "New Species in the Old World: Europe as a Frontier in Biodiversity Exploration, A Test Bed for 21st Century Taxonomy," *PLOS One* 7, no. 5 (2012): e36881, https://doi.org/10.1371/journal.pone.0036881; S. Brusatte, *The Rise and Fall of the Dinosaurs: A New History of a Lost World* (New York: HarperCollins, 2018).

OUR GENEALOGY AND ANCESTRY

1. K. Mack, *The End of Everything: Astrophysics and the Ultimate Fate of the Cosmos* (New York: Scribner, 2020); C. Prescod-Weinstein, *The Disordered Cosmos: A Journey into Dark Matter, Spacetime, and Dreams Deferred* (London: Hachette UK, 2021).

2. P. Edmonds, "These Twins Will Make You Rethink Race," *National Geographic*, April 2018 https://www.nationalgeographic.com/magazine/2018/04/race-twins-black-white-biggs/.

3. J. Woodson, *Brown Girl Dreaming* (New York: Penguin Random House, 2016).

4. G. Coop, "How Much of Your Genome Do You Inherit from a Particular Grandparent?" *Gcbias*, posted October 20, 2013, https://gcbias.org/2013/10/20/how-much-of-your-genome-do-you-inherit-from-a-particular-grandparent/.

5. Coop, "How Much of Your Genome."

6. D. Enard and D. A. Petrov, "Evidence That RNA Viruses Drove Adaptive Introgression between Neanderthals and Modern Humans,"

Cell 175, no. 2 (2018): 360–371.e13, https://doi.org/10.1016/j.cell
.2018.08.034; P. Chakrabarty, "What Can DNA Tests Really Tell Us
about Our Ancestry?" TED-Ed, June 9, 2020, https://www.ted.com/talks
/prosanta_chakrabarty_what_can_dna_tests_really_tell_us_about_our
_ancestry?language=en.

7. Enard and Petrov, "Evidence That RNA Viruses Drove."

8. S. Gazal, M. Sahbatou, M. C. Babron, E. Génin, and A. L. Leute-
negger, "High Level of Inbreeding in Final Phase of 1000 Genomes
Project," *Scientific Reports* 5, article 17453 (2015).

9. A. Horton, "Elizabeth Warren Angers Prominent Native Americans
with Politically Fraught DNA Test," *Washington Post*, October 16, 2018,
https://www.washingtonpost.com/nation/2018/10/16/elizabeth
-warren-angers-prominent-native-americans-with-politically-fraught
-dna-test/.

10. N. A. Garrison, "Genetic Ancestry Testing with Tribes: Ethics,
Identity and Health Implications," *Daedalus* 147, no. 2 (2018): 60–69,
https://doi.org/10.1162/DAED_a_00490.

11. C. Hilleary, "Native Americans Speak Out on Elizabeth Warren
DNA Controversy," *Voice of America*, October 16, 2018, https://www
.voanews.com/a/native-americans-speak-out-on-elizabeth-warren
-dna-controversy/4615743.html?fbclid=IwAR2WaKcxC3G52su9Ancv
ZTHRj35RY3vAgl4-7tAt8I0xZyCjeixxe77Ae0A.

12. J. K. Wagner, "Interpreting the Implications of DNA Ancestry
Ttests," *Perspectives in Biology and Medicine* 53, no. 2 (2010): 231–248,
https://doi.org/10.1353/pbm.0.0158.

13. A. Auton, L. D. Brooks, R. M. Durbin, E. P. Garrison, H. M. Kang, J.
O. Korbel, et al., "A Global Reference for Human Genetic Variation,"
Nature 526, no. 7571 (2015): 68–74, https://www.ncbi.nlm.nih.gov
/pmc/articles/PMC4750478/.

14. Auton et al., "Global Reference."

15. "Ancestry Composition: 23andMe's State-of-the-Art Geographic
Ancestry Analysis," 23andMe, accessed October 8, 2022, https://www
.23andme.com/ancestry-composition-guide/.

16. H. M. Cann, C. de Toma, L. Cazes, M.-F. Legrand, V. Morel, K.
Pioffre, et al., "A Human Genome Diversity Cell Line Panel," *Science*
296, no. 5566 (2002): 261–262, https://doi.org/10.1126/science.296

.5566.261b; 1000 Genomes Project Consortium, G. R. Abecasis, D. Altshuler, A. Auton, L. D. Brooks, R. M. Durbin, et al., "A Map of Human Genome Variation from Population-Scale Sequencing," *Nature* 467, no. 7319 (2010): 1061–1073, https://doi.org/10.1038/nature09534.

17. A. Hamon, "Why White Supremacists Are Chugging Milk (and Why Geneticists Are Alarmed), *New York Times*, October 17, 2018, https://www.nytimes.com/2018/10/17/us/white-supremacists-science -dna.html.

18. A. Hamon, "Geneticists Criticize Use of Science by White Nationalists to Justify 'Racial Purity,'" *New York Times*, October 19, 2018, https://www.nytimes.com/2018/10/19/us/white-supremacists-science -genetics.html.

19. V. Stănescu, "'White Power Milk': Milk, Dietary Racism, and the 'Alt-Right,'" *Animal Studies Journal* 7, no. 2 (2018): 103–128.

20. P. Singh, A. Arora, T. A. Strand, D. A. Leffler, C. Catassi, P. H. Green, et al., "Global Prevalence of Celiac Disease: Systematic Review and Meta-Analysis," *Clinical Gastroenterology and Hepatology* 16, no. 6 (2018): 823–836.e2, https://doi.org/10.1016/j.cgh.2017.06.037.

21. Stănescu, "'White Power Milk.'"

22. P. Gerbault, A. Liebert, Y. Itan, A. Powell, M. Currat, J. Burger, et al., "Evolution of Lactase Persistence: An Example of Human Niche Construction," *Philosophical Transactions of the Royal Society B: Biological Sciences* 366, no. 1566 (2011): 863–877, https://doi.org/10.1098 /rstb.2010.0268.

23. Gerbault et al., "Evolution of Lactase Persistence."

24. American Society of Human Genetics, "ASHG Denounces Attempts to Link Genetics and Racial Supremacy," *American Journal of Human Genetics* 103, no. 5 (2018): 636, https://doi.org/10.1016/j.ajhg .2018.10.011; K. Dunlap, "The Great Aryan Myth," *Scientific Monthly* 59, no. 4 (1944): 296–300.

25. A. Saini, *Superior: The Return of Race Science* (Boston: Beacon Press, 2019); Lewontin, R. C., "The Apportionment of Human Diversity," in *Evolutionary Biology*, ed. T. Dobzhansky, M. K. Hecht, and W. C. Steer (New York: Springer, 1972), 381–398; M. Yudell, D. Roberts, R. DeSalle, and S. Tishkoff, "Taking Race Out of Human Genetics," *Science* 351, no. 6273 (2016): 564–565, https://doi.org/10.1126/science .aac4951.

26. R. S. Zimmerman, W. A. Vega, A. G. Gil, G. J. Warheit, E. Apospori, and F. Biafora, "Who Is Hispanic? Definitions and Their Consequences," *American Journal of Public Health* 84, no. 12 (1994): 1985–1987, https://doi.org/10.2105/AJPH.84.12.1985.

27. C. Salinas Jr. and A. Lozano, "Mapping and Recontextualizing the Evolution of the Term *Latinx*: An Environmental Scanning in Higher Education," *Journal of Latinos and Education* 18, no. 4 (2019): 302–315, https://doi.org/10.1080/15348431.2017.1390464.

28. Saini, *Superior*; D. N. Lee, "Diversity and Inclusion Activisms in Animal Behaviour and the ABS: A Historical View from the U.S.A.," *Animal Behaviour* 164 (2020): 273–280, https://doi.org/10.1016/j .anbehav.2020.03.019; J. L. Graves, "African Americans in Evolutionary Science: Where We Have Been, and What's Next," *Evolution: Education and Outreach* 12, article 18 (2019), https://doi.org/10.1186 /s12052-019-0110-5; P. H. Barber, T. B. Hayes, T. L. Johnson, and L. Márquez-Magaña, "Systemic Racism in Higher Education," *Science* 369, no. 6510 (2020): 1440–1441; K. G. Claw, M. Z. Anderson, R. L. Begay, K. S. Tsosie, K. Fox, N. A. Garrison, et al., "A Framework for Enhancing Ethical Genomic Research with Indigenous Communities," *Nature Communications* 9, article 2957 (2018), https://doi .org/10.1038/s41467-018-05188-3.

29. S. Baharian, M. Barakatt, C. R. Gignoux, S. Shringarpure, J. Errington, W. J. Blot, et al., "The Great Migration and African–American Genomic Diversity," *PLOS Genetics* 12, no. 5 (2016): e1006059, https:// doi.org/10.1371/journal.pgen.1006059.

30. J. Lelyveld, *Great Soul: Mahatma Gandhi and His Struggle with India* (New York: Vintage Books/Random House, 2012); F. J. Davis, *Who Is Black? One Nation's Definition* (University Park: Pennsylvania State University Press, 1991); L. L. Martin, *Racial Realism and the History of Black People in America* (Washington, DC: Rowman & Littlefield, 2022).

31. Barack Obama, *Dreams from My Father: A Story of Race and Inheritance* (Edinburgh: Canongate Books, 2007).

32. F. J. Davis, *Who Is Black? One Nation's Definition* (University Park: Pennsylvania State University Press, 1991).

33. J. Daniszewski, "AP Stylebook Updates Race-Related Terms," *ACES* (blog), posted February 2, 2021, https://aceseditors.org/news/2021/ap -stylebook-updates-race-related-terms.

34. J. L. Graves Jr., "Why the Nonexistence of Biological Races Does Not Mean the Nonexistence of Racism," *American Behavioral Scientist* 59, no. 11 (2015): 1474–1495.

SEX/GENDER/SEXUALITY

1. Y. Barazani and E. Sabanegh Jr., "Rare Case of Monozygotic Twins Diagnosed with Klinefelter Syndrome during Evaluation for Infertility," *Nature Reviews Urology* 17, no. 1 (2015): 42–45.

2. A. L. Reiss, L. Freund, L. Plotnick, T. Baumgardner, K. Green, A. C. Sozer, et al., "The Effects of X Monosomy on Brain Development: Monozygotic Twins Discorcant for Turner's Syndrome," *Annals of Neurology* 34, no. 1 (1993): 95–107, https://doi.org/10.1002/ana.410 340117; M. E. De Paepe, "Multiple Gestation: The Biology of Twinning," in *Creasy and Resnik's Maternal-Fetal Medicine: Principles and Practice*, 8th ed., ed. R. Resnik, C. J. Lockwood, T. R. Moore, et al. (Philadelphia: Elsevier, 2018), 68–80.

3. A. Montañez, "Beyond XX and XY," *Scientific American* 317, no. 3 (March 2017): 50–51, https://doi.org/10.1038/scientificamerican0917-50.

4. I. A. Hughes, J. D. Davies, T. I. Bunch, V. Pasterski, K. Mastroyannopoulou, and J. MacDougall, "Androgen Insensitivity Syndrome," *Lancet* 380, no. 9851 (2012): 1419–1428.

5. Montañez, "Beyond XX and XY."

6. A. Zeiher and T. Braun, "Mosaic Loss of Y Chromosome during Aging," *Science* 377, no. 6603 (2022): 266–267.

7. A. Fausto-Sterling, *Sexing the Body: Gender Politics and the Construction of Sexuality* (Basic Books, 2000).

8. Fausto-Sterling, *Sexing the Body*.

9. F. J. Janzen and P. C. Phillips, "Exploring the Evolution of Environmental Sex Determination, Especially in Reptiles," *Journal of Evolutionary Biology* 19, no. 6 (2006): 1775–1784.

10. J. Roughgarden, *Evolution's Rainbow: Diversity, Gender, and Sexuality in Nature and People* (Berkeley: University of California Press, 2009).

11. D. Gonçalves, R. F. Oliveira, K. Körner, and I. Schlupp, "Intersexual Copying by Sneaker Males of the Peacock Blenny," *Animal*

Behaviour 65, no. 2 (2003): 355–361, https://doi.org/10.1006/anbe .2003.2065.

12. R. T. Hanlon, M.-J. Naud, P. W. Shaw, and J. N. Havenhand, "Transient Sexual Mimicry Leads to Fertilization," *Nature* 433, no. 212 (2005), https://doi.org/10.1038/433212a.

13. L. Casas, F. Saborido-Rey, T. Ryu, C. Michell, T. Ravasi, and X. Irigoien, "Sex Change in Clownfish: Molecular Insights from Transcriptome Analysis," *Scientific Reports* 6, article 35461 (2016), https:// doi.org/10.1038/srep35461.

14. J. Imperato-McGinley, L. Guerrero, T. Gautier, and R. E. Peterson, "Steroid 5alpha-Reductase Deficiency in Man: An Inherited Form of Male Pseudohermaphroditism," *Science* 186, no. 4170 (1974): 1213–1215, https://doi.org/10.1126/science.186.4170.1213.

15. A. E. Nutt, *Becoming Nicole* (New York: Penguin Random House, 2016).

16. M. G. Peletz, *Gender Pluralism: Southeast Asia Since Early Modern Times* (New York: Routledge, 2009); S. E. Jacobs, W. Thomas, and S. Lang, *Two-Spirit People: Native American Gender Identity, Sexuality, and Spirituality* (Urbana: University of Illinois Press, 1997).

17. J. T. Bell and T. D. Spector, "DNA Methylation Studies Using Twins: What Are They Telling Us?" *Genome Biology* 13, no. 10 (2012): 172, https://doi.org/10.1186/gb-2012-13-10-172.

18. M. Diamond, "Transsexuality among Twins: Identity Concordance, Transition, Rearing, and Orientation," *International Journal of Transgenderism* 14, no. 1 (2013): 24–38, https://doi.org/10.1080/1553 2739.2013.750222; T. C. Ngun and E. Vilain, "The Biological Basis of Human Sexual Orientation," *Advances in Genetics* 86 (2014): 167–184, https://doi.org/10.1016/b978-0-12-800222-3.00008-5; A. Ganna, K. J. H. Verweij, M. G. Nivard, R. Maier, R. Wedow, A. S. Busch, et al., "Large-Scale GWAS Reveals Insights into the Genetic Architecture of Same-Sex Sexual Behavior," *Science* 365, no. 6456 (2019): eaat7693.

19. T. C. Ngun and E. Vilain, "The Biological Basis of Human Sexual Orientation," *Advances in Genetics* 86 (2014): 167–184, https://doi.org /10.1016/b978-0-12-800222-3.00008-5; Ganna et al., "Large-Scale GWAS Reveals Insights."

20. Diamond, "Transsexuality among Twins."

21. Diamond, "Transsexuality among Twins."

22. V. Sommer and P. L. Vasey, *Homosexual Behaviour in Animals: An Evolutionary Perspective* (Cambridge: Cambridge University Press, 2006); B. Bagemihl, *Exuberance: Animal Homosexuality and Natural Diversity* (New York: St. Martin's Press, 2000).

23. J. Roughgarden, *Evolution's Rainbow: Diversity, Gender, and Sexuality in Nature and People* (Berkeley: University of California Press, 2009); Sommer and Vasey, *Homosexual Behaviour in Animals*.

24. C. Bull, "Birds of a Feather: Meet Wendell and Cass, the Gay Male Penguin Couple at the New York Aquarium," *Advocate* magazine. April 2, 2002; K. S. Ebeling and B. B. Spanier, "What Made Those Penguins Gay?" in *Gender and the Science of Difference: Cultural Politics of Contemporary Science and Medicine*, ed. J. A. Fisher (New Brunswick: Rutgers University Press, 2011), 126–144.

25. J. Yoder, "A New Age of Gay Genomics Is Here. Are We Ready for the Consequences?" *Slate*, August 29, 2019, https://slate.com/tech nology/2019/08/gay-gene-study-ganna-biobank-23andme-risks.html.

26. Ganna et al., "Large-Scale GWAS Reveals Insights."

27. T. Burki, "Genetic Apps: Raising More Questions Than They Answer?" *Lancet Digital Health* 2, no. 1 (2020): 13–14, https://www .thelancet.com/journals/landig/issue/vol2no1/PIIS2589-7500(19) X0010-2.

28. W. D. Hill, N. M. Davies, S. J. Ritchie, N. G. Skene, J. Bryois, S. Bell, et al., "Genome-Wide Analysis Identifies Molecular Systems and 149 Genetic Loci Associated with Income," *Nature Communications* 10, article 5741 (2019), https://doi.org/10.1038/s41467-019-13585-5.

29. J. J. Lee, R. Wedow, A. Okbay, E. Kong, O. Maghzian, M. Zacher, et al., "Gene Discovery and Polygenic Prediction from a Genome-Wide Association Study of Educational Attainment in 1.1 Million Individuals," *Nature Genetics* 50 (2018): 1112–1121, https://doi.org/10.1038/s 41588-018-0147-3.

30. E. Hall, *Aristotle's Way: How Ancient Wisdom Can Change Your Life* (New York: Penguin Random House, 2019).

31. K. Keely, "Poverty, Sterilization, and Eugenics in Erskine Caldwell's *Tobacco Road*," *Journal of American Studies* 36, no. 1 (2002): 23–42, https://doi.org/10.1017/S002187580200676X.

32. E. Cultotta, "Probing an Evolutionary Riddle," *Science* 365, no. 6455 (2019): 748–749.

33. J. D. Monk, E. Giglio, A. Kamath, M. R. Lambert, and C. E. McDonough, "An Alternative Hypothesis for the Evolution of Same-Sex Sexual Behaviour in Animals," *Nature: Ecology & Evolution* 3 (2019): 1622–1651.

34. J. M. Jones, "LGBT Identification in U.S. ticks up to 7.1%," Gallup, February 2022, https://news.gallup.com/poll/389792/lgbt-identification-ticks-up.aspx; G. J. Gates, "How Many People Are Gay, Lesbian, Bisexual and Transgender?" UCLA Williams Institute Law School, April 2011, https://williamsinstitute.law.ucla.edu/publications/how-many-people-lgbt/; J. McKnight, *Straight Science? Homosexuality, Evolution and Adaptation* (London: Routledge, 1997).

35. T. Casci, "Gay Genes Boost Fertility," *Nature Reviews Genetics* 5 (2004): 884, https://doi.org/10.1038/nrg1510.

36. J. McKnight and J. Malcolm, "Is Male Homosexuality Maternally Linked?" *Psychology, Evolution & Gender* 2 no. 3 (2000): 229–252, https://doi.org/10.1080/14616660010024599.

37. McKnight and Malcolm, "Is Male Homosexuality Maternally Linked?'

38. E. O. Wilson, *Sociobiology: The New Synthesis, Twenty-Fifth Anniversary Edition* (Belknap Press/Harvard University Press, 1975); J. Roughgarden, "Homosexuality and Evolution," in *On Human Nature: Biology, Psychology, Ethics, Politics, and Religion*, ed. M. Tibayrenc and F. J. Ayala (London: Academic Press, 2017), 495–516, https://doi.org/10.1016/b978-0-12-420190-3.00030-2.

39. S. Jones, A. J. Webster, and A. T. Zemnick, "Inclusive and Accurate Methods for Teaching Sex- and Gender-Related Topics" (Google slides), Project Biodiversity, 2019, https://projectbiodiversify.org/sex/.

40. Darwin, *On the Origin of Species* (1859 edition); Darwin, *Descent of Man*.

41. Darwin, *Descent of Man*.

42. C. Wilcox, "Males Are the Taller Sex. Estrogen, Not Fights for Mates, May Be Why," *Quanta* magazine, June 8, 2020; H. M. Dunsworth, "Expanding the Evolutionary Explanations for Sex Differences

in the Human Skeleton," *Evolutionary Anthropology: Issues, News, and Reviews* 29, no. 3 (2020): 108–116.

43. H. Kokko, M. D. Jennions, and R. Brooks, "Unifying and Testing Models of Sexual Selection," *Annual Review of Ecology, Evolution, and Systematics* 37 (2006): 43–66; M. Andersson, *Sexual Selection*, vol. 72 (Princeton: Princeton University Press, 1994).

COMBATTING POST-TRUTH WITH TRUST

1. T. H. Swartz, A. G. S. Palermo, S. K. Masur, and J. A. Aberg, "The Science and Value of Diversity: Closing the Gaps in Our Understanding of Inclusion and Diversity," *Journal of Infectious Diseases* 220, suppl. 2 (2019): S33–S41.

2. A. J. Reid, L. E. Eckert, J. F. Lane, N. Young, S. G. Hinch, C. T. Darimont, et al., "'Two-Eyed Seeing': An Indigenous Framework to Transform Fisheries Research and Management," *Fish and Fisheries* 22 (2020): 243–261; Roy, R. D., "Decolonise Science—Time to End Another Imperial Era," The Conversation (website), April 5, 2018, https://theconversation.com/decolonise-science-time-to-end-another -imperial-era-89189.

3. L. Lundgren, K. Stofer, B. Dunckel, J. Krieger, M. Lange, and V. James, "Panel-Based Exhibit Using Participatory Design Elements May Motivate Behavior Change," *Journal of Science Communication* 18, no. 2 (2019): A03; F. Heigl, B. Kieslinger, K. T. Paul, J. Uhlik, and D. Dörler, "Opinion: Toward an International Definition of Citizen Science," *Proceedings of the National Academy of Sciences* 116, no. 17 (2019): 8089–8092, https://doi.org/10.1073/pnas.1903393116; M. Aristeidou and D. Herodotou, "Online Citizen Science: A Systematic Review of Effects on Learning and Scientific iteracy," *Citizen Science: Theory and Practice* 5, no. 1, article 11 (2020).

CONVERSATION WITH A CREATIONIST

1. L. Pray, "Eukaryotic Genome Complexity," *Nature Education* 1, no. 1 (2008): 96; H. Ashworth, "How Long Is Your DNA?" *BBC Science Focus* magazine, 2020, https://www.sciencefocus.com/the-human-body /how-long-is-your-dna/.

2. K. Vonnegut, *Galápagos: A Novel* (New York: Dial Press, 2009).

EPILOGUE

1. Donald J. Trump, in J. S. Brady, "Remarks by President Trump, Vice President Pence, and Members of the Coronavirus Task Force in Press Briefing," *Whitehouse.gov*, April 1, 2020, https://trumpwhitehouse .archives.gov/briefings-statements/remarks-president-trump-vice -president-pence-members-coronavirus-task-force-press-briefing-16/.

2. D. A. Kennedy and A. F. Read, "Monitor for COVID-19 Vaccine Resistance Evolution during Clinical Trials," *PLOS Biology* 18, no. 11 (2020): e3001000.

3. Kennedy and Read, "Monitor for COVID-19 Vaccine Resistance."

4. H. Kalish, C. Klumpp-Thomas, S. Hunsberger, H. A. Baus, M. P. Fay, N. Siripong, et al., "Undiagnosed SARS-CoV-2 Seropositivity during the First Six Months of the COVID-19 Pandemic in the United States," *Science Translational Medicine* 13, no. 601 (2021): eabh3826.

5. A. V. Raveendran, R. Jayadevan, and S. Sashidharan, "Long COVID: An Overview," *Diabetes & Metabolic Syndrome: Clinical Research & Reviews* 15, no. 3 (2021): 869–875; J. Couzin-Frankel, "Clues to Long COVID," *Science* 376, no. 6599 (2022): 1261–1265.

6. B. Wasik and M. Murphy, *Rabid: A Cultural History of the World's Most Diabolical Virus* (New York: Penguin Books, 2012).

7. P. Mehta, D. F. McAuley, M. Brown, E. Sanchez, R. S. Tattersall, and J. J. Manson, "COVID-19: Consider Cytokine Storm Syndromes and Immunosuppression," *Lancet* 395, no. 10229 (2020): 1033–1034, https://doi.org/10.1016/S0140-6736(20)30628-0.

8. L. Spinney, *Pale Rider: The Spanish Flu of 1918 and How It Changed the World* (New York: Public Affairs, 2017).

9. S. Shah, *Pandemic: Tracking Contagions from Cholera to Ebola and Beyond* (New York: Farrar, Straus and Giroux, 2016); C. Millard, *Destiny of the Republic: A Tale of Madness, Medicine and the Murder of a President* (New York: Anchor Books/Doubleday, 2011).

10. C. Zimmer, "Most New York Coronavirus Cases Came from Europe, Genomes Show," *New York Times*, April 8, 2020, https://www .nytimes.com/2020/04/08/science/new-york-coronavirus-cases-europe -genomes.html.

11. K. Sadtler and E. Wayne, "How the COVID-19 Vaccines Were Created So Quickly," TED-Ed, August 17, 2021, https://www.ted.com/talks

/kaitlyn_sadtler_and_elizabeth_wayne_how_the_covid_19_vaccines _were_created_so_quickly?language=en; L. A. Jackson, E. J. Anderson, N. G. Rouphael, P. C. Roberts, M. Makhene, R. N. Coler, et al., "An mRNA Vaccine against SARS-CoV-2—Preliminary Report," *New England Journal of Medicine* 383 (2020): 1920–1931.

12. S. Mehta, "Trump's Touting of 'Racehorse Theory' Tied to Eugenics and Nazis Alarms Jewish Leaders," *Los Angeles Times*, October 9, 2020, https://www.latimes.com/politics/story/2020-10-05/trump-debate -white-supremacy-racehorse-theory; N. Wu, "Trump Criticized for Praising 'Good Bloodlines' of Henry Ford, Who Promoted Anti-Semitism," *USA Today*, May 22, 2020, https://www.usatoday.com/story /news/politics/2020/05/22/trump-criticized-praising-bloodlines-henry -ford-anti-semite/5242361002/.

13. C. Sohini, "Bowel Cleanse for Better DNA: The Nonsense Science of Modi's India," *This Week in Asia*, January 13, 2019, https:// www.scmp.com/week-asia/society/article/2181752/bowel-cleanse -better-dna-nonsense-science-modis-india.

GLOSSARY

1. E. Waltz, "Nonbrowning GM Apple Cleared for Market," *Nature Biotechnology* 33, no. 4 (2015): 326–328.

2. J. Hutton, *Theory of the Earth: With Proofs and Illustrations*, vol. 1 (Edinburgh, 1795); Lyell, *Principles of Geology*, vol. 1.

3. Genesis (King James Version); R. L. Numbers, *The Creationists: From Scientific Creationism to Intelligent Design* (Cambridge, MA: Harvard University Press, 2006).

4. Numbers, *The Creationists*; C. Funk, L. Rainie, and D. Page, *Americans, Politics and Science Issues* (Washington, DC: Pew Research Center, July 1, 2015), 1.

5. D. W. Deamer and G. Fleischaker, *Origins of Life: The Central Concepts* (Sudbury, MA: Jones & Bartlett, 1994).

6. H. M. B. Harris and C. Hill, "A Place for Viruses on the Tree of Life," *Frontiers in Microbiology* 11 (2021): 604048.

FURTHER READING

Learn more about the science of evolution from Riley Black, Deborah Blum, Laura Boykin, Sean B. Carroll, Richard Conniff, Jonathan Eisen, Keolu Fox, Douglas Futuyma, Raghavendra Gadagkar, Amanda L. Glaze Townley, Emily Graslie, Joseph L. Graves Jr., Hank Green, Hopi Hoekstra, Danielle N. Lee, Richard Lenski, Jonathan Losos, Ken R. Miller, P. Z. Myers, Mohamed Noor, C. Brandon Ogbunu, David Quammen, Andrea Reid, Leslie Rissler, Anna Rothschild, Helen Scales, Eugenie Scott, Neil Shubin, Scott E. Solomon, Jessica Ware, Mary Jane West-Eberhard, Christie Wilcox, Katherine Wu, Jeremy Yoder, Ed Yong, Carl Zimmer, and many other wonderful scientists, journalists, explorers, and thinkers.

FIGURE CREDITS

Cover Art Image by Akangksha Sarmah.

Figures 1A and 1B Images from Shutterstock and T. Michael Keesey (PhyloPic.org), modified by Prosanta Chakrabarty.

Figure 2/plate 1 Image by Prosanta Chakrabarty with silhouettes from PhyloPic.org. Shark—no copyright, http://phylopic.org/image /545d45f0-0dd1-4cfd-aad6-2b835223ea0d/; fish—http://phylopic.org /image/f1f91d08-b850-4600-ad64-622ce87f0199/; seahorse—http://phy lopic.org/image/1ce0d9a9-c644-4cc6-b4b3-02e2e7aa2266/.

Figure 3 Image by Sumantra Mukherjee.

Figure 4/plate 2 Modified from L. A. Hug, B. J. Baker, K. Ananthara-man, C. T. Brown, A. J. Probst, C. J. Castelle, et al., "A New View of the Tree of Life," *Nature Microbiology* 1, no. 11 (April 2016), http://doi.org /10.1038/nmicrobiol.2016.48. Free to reprint—https://creativecommons .org/licenses/by/4.0/ (B) simplified tree from Wikipedia https://en .wikipedia.org/wiki/Phylogenetic_tree.

Figure 5 Image by Sumantra Mukherjee.

Figures 6A and 6B/plates 3 and 4 Image by Prosanta Chakrabarty. Portions of silhouette images from T. Michael Keesey and PhyloPic .org.

Figure 7/plate 5 Image by Sumantra Mukherjee.

Figure 8/plate 6 Image by Sumantra Mukherjee.

Figure 9 Image by Sumantra Mukherjee.

Figure 10 Image by Sumantra Mukherjee.

Figure 11/plate 7 Image by Prosanta Chakrabarty.

Figure 12 Image by Sumantra Mukherjee.

Figure 13 Creative Commons 3: Reprinted from I. Braasch, A. Gehrke, J. Smith, J., K. Kawasaki, T. Masousaki, J. Pasquier, et al., "The Spotted Gar Genome Illuminates Vertebrate Evolution and Facilitates Human-Teleost Comparisons," *Nature Genetics* 48 (2016): 427–437, https://doi.org/10.1038/ng.3526.

Figure 14 Photo by Prosanta Chakrabarty.

Darwin movie panels All illustrations created by Ethan Kocak for this book.

Prosanta Chakrabarty is the George H. Lowery Jr. Professor and Curator of Fishes at the Museum of Natural Science and Department of Biological Sciences at Louisiana State University in Baton Rouge. He is also a research associate at the American Museum of Natural History, Smithsonian's National Museum of Natural History, and the Canadian Museum of Nature. He is a systematist and an ichthyologist studying the evolution and biogeography of both freshwater and marine fishes. He has described over a dozen new species to science and published more than one hundred scientific papers. He grew up in New York City; his undergraduate degree is from McGill University in Montreal, the city where he was born, and his PhD is from the University of Michigan. He is a former Program Director at the National Science Foundation, an elected fellow of the American Association for the Advancement of Science, a TED Senior Fellow, and a Fulbright Distinguished Chair. He teaches Evolution and Ichthyology at LSU where he is also the Chair of the Center for Collaborative Knowledge. He is the father of eleven-year-old twins and has been married to his wife, Annemarie, for eighteen years. He loves them a lot but thought dedicating this book to them would be a bit much.

You can follow him on Twitter @PREAUX_FISH or his Facebook page "Dr. Prosanta Chakrabarty."

INDEX